THE OPEN UNIVERSITY

Science:
A Second Level Course

S299
GENETICS

14 and 15 Human Genetics

Prepared by a Course Team for the Open University

THE OPEN UNIVERSITY PRESS

Course Team

Chairman and General Editor
Steven Rose

Unit Authors
Norman Cohen (*The Open University*)
Terence Crawford-Sidebotham (*University of York*)*
Denis Gartside (*University of Hull*)
David Jones (*University of Hull*)
Steven Rose (*The Open University*)
Derek Smith (*University of Birmingham*)
Mike Tribe (*University of Sussex*)
Robert Whittle (*University of Sussex*)

*Consultant

Editor
Jacqueline Stewart

Other Members
Bob Cordell (*Staff Tutor*)
Mae-Wan Ho*
Jean Holley (*Technician*)
Stephen Hurry
Roger Jones (*BBC*)
Aileen Llewellyn (*Course Assistant*)
Michael MacDonald-Ross (*IET*)
Jean Nunn (*BBC*)
Pat O'Callaghan (*Evaluation*)
Jim Stevenson (*BBC*)

* From January 1976

The development of this Course was supported by a grant from the Nuffield Foundation.

The Open University Press,
Walton Hall, Milton Keynes.

First published 1976. Reprinted 1979.

Designed by the Media Development Group of the Open University.
Set by Composition House Ltd, Salisbury, Wiltshire.

Printed in Great Britain by Eyre and Spottiswoode Limited,
at Grosvenor Press, Portsmouth.

ISBN 0 335 04289 9

This text forms part of an Open University Course. The complete list of Units in the Course appears at the end of this text.

For general availability of supporting material referred to in this text please write to the Director of Marketing, The Open University, P.O. Box 81, Walton Hall, Milton Keynes, MK7 6AT.

Further information on Open University Courses may be obtained from the Admissions Office, The Open University, P.O. Box 48, Walton Hall, Milton Keynes, MK7 6AB.

1.2

14 and 15 Human Genetics

Contents

List of scientific terms used in Units 14 and 15

Developed in this Unit	Page No.	Developed in this Unit	Page No.
amniocentesis	673	'hereditarian' thesis	659
behavioural genetics	641	hybrid cells	631
behavioural mutant	642	inbreeding	671
behavioural phenotype	641	IQ	660
between-group differences	666	killer rats	643
biological and social definition of race	651, 657	kinship correlations	663
		learning ability	643
breeding structure	671	maze-bright, maze-dull	643
cell clone	630	mental retardation	645
cell line	630	monozygotic (MZ) twins	626
codominant allele	638		
concordance	627	natural selection in human populations	668
consanguineous mating	672	norm of reaction	670
dizygotic (DZ) twins	626	partial penetrance	624
electrophoretic identity	629	phenotypic similarity	623
environment	642	phenylketonuria	623
eugenics	675	phylogeny	652
euphenics	646	polygenic control	625
evolutionary divergence	652	polytypic species	657
galactosaemia	674	rhesus positive, rhesus negative	636
gene flow	656	schizophrenia	647
genetic counselling	673	separated twins	664
genetic disease	635	somatic-cell genetics	630
genetic engineering	677	threshold effects	625
genetic variation in humans	648	within-group differences	666
haemolytic disease	636		

Objectives for Units 14 and 15

After studying these Units, you should be able to:

1   List and discuss the constraints of genetic methodology when applied to human genetics.
(ITQ 18)

2   Recognizing the problem of ascertainment, construct genetic hypotheses from pedigree data and population sampling.
(ITQs 1, 2, 3, 15 and 16)

3   Describe, giving specific examples, the application of somatic-cell genetics and cell fusion to genetic analysis in humans.
(ITQs 6 and 7)

4   Review the advantages and shortcomings of using twins to investigate the genetic components of variation.
(ITQs 4 and 5)

5   Contrast the approaches and information gained from pedigree analysis, studies of twins and somatic-cell genetics.
(SAQs 1 and 5)

6   Discuss the contribution made by genetics to the understanding and treatment of rhesus haemolytic disease.
(ITQs 8, 9, 10 and 11)

7   Analyse the particular problems of studying behaviour as a phenotype in human and non-human species.
(ITQs 12, 13 and 14)

8   Review the contribution of genetics to the understanding and treatment of schizophrenia.
(ITQ 17; SAQ 5)

9   Distinguish between biological and social concepts of race.
(ITQ 19; SAQ 2)

10   Construct phylogenies from allele-frequency data and recall the assumptions made in doing so.
(ITQs 20, 21, 24 and 30)

11   Review the possible explanations for the genetic diversity within and between local populations.
(ITQs 21, 22, 23 and 24)

12   Discuss the problems of measuring 'IQ' and defining it as a distinct phenotype.
(ITQs 25 and 26; SAQs 3 and 4)

13   Review the evidence relevant to the hypothesis that there is genetic variation affecting IQ.
(SAQ 5)

14   Discuss the significance and limitations of heritability estimates as applied to human characters within and between groups.
(SAQ 6)

15   Consider the potential effectiveness of eugenic measures against the possibility of deleterious recessive and dominant mutations, and the complications of incomplete penetrance.
(ITQs 26, 27, 28 and 29)

16   Review the impact on the gene pool of attempts to cure/normalize/support individuals with 'genetic diseases'.
(ITQs 26, 27, 28, 29, 30, 31, 32, 33 and 34; SAQ 8)

17   Distinguish between 'genetic counselling', euphenics, 'genetic engineering' and eugenics. Give examples of actual or proposed developments in each, and discuss their practicability and likely outcomes.
(ITQ 34; SAQ 8)

18   Consider the social circumstances in which arguments about eugenic proposals and genetic differences between social groups may be expected to occur.
(SAQs 7 and 8)

Study guide for Units 14 and 15

These two Units on human genetics are not only the last of the Course but also in a way its culmination—the attempt to use the principles, methods and theories established in the previous Units to explain and predict the attributes and behaviour of human individuals and populations. These topics are not 'purely academic' (if ever there was such a thing); they have been and are controversial, raising issues of intense human interest and practical social and medical significance. In dealing with them we are not necessarily neutral—though we do claim to be 'objective'! We want you to be aware of the issues and how they affect you/society, and we hope you will be interested enough to extend your reading to include some of the books referred to in these Units.

So far as the structure of the double Unit is concerned, we believe it to be about two Units worth in overall study time, but the break in the text between Units 14 and 15 is fairly arbitrary. There are no major new genetic concepts introduced in the Units; instead they are intended to show the particular strengths and problems of applying the genetic methods discussed early in the Course to humans. The double Unit begins with a Section (Section 14.1) on the particular problems of applying genetic techniques to the study of humans; this Section is vital for what comes later, and you must achieve a thorough grasp of it (Objectives 1–5). There follows the first of the two major 'case studies' of the Units (Section 14.2), on rhesus haemolytic disease (Objective 6). If you feel yourself short of time, do not attempt to learn this material—its details will *not* be assessed.

The flow of the discussion of human genetics is then broken to introduce a new topic—behavioural genetics (Section 14.3). However, before turning to the study of humans, the behavioural genetics of non-human animals is discussed so as to reveal the particular problems surrounding the study of behaviour as a phenotype (Objectives 7 and 8). Objective 8 should be omitted if you are short of time.

Unit 15 begins with a discussion of human population genetics and distinguishes between the biological and social uses of the term 'race'. Objectives 9–11 are important and should on no account be neglected.

The material so far presented makes possible an approach to the second of the major case studies of these Units (Section 15.3, Objectives 12–14)—on the question of intelligence and the so-called 'race–IQ' debate, perhaps the most sensitive area of the entire Course. You are encouraged to study this material, to try to look at the additional recommended reading associated with it, and above all to assimilate the more general genetic points made at the end of the case study (Sections 15.3.6–15.3.8) in relation to Objective 14.

Finally, in Section 15.4.5, the question of the continuing evolution of human populations and the implications of various 'genetic engineering', 'genetic counselling' and eugenic proposals are discussed (Objectives 15–18). It is important that you read and understand this material.

There are three TV programmes associated with these Units: TV programme 14 deals with molecular evolution, TV programme 15 with human diversity and TV programme 16 with genetic engineering, primarily in plants.

The associated radio programmes are Radio programme 14 (genetic counselling) and Radio programme 16 (genetics and society).

Introduction to Units 14 and 15: why study human genetics?

> When I was about eight years old, my father took me to the American Museum of Natural History and, as I well remember, made clear to me, through the simple example of the succession of fossil horses' feet shown there, how organs and organisms became gradually changed through the interaction of accidental variation and natural selection ... From that time the idea never left my head that if this could happen in nature, men should eventually be able to control the process, even in themselves, so as greatly to improve upon their own natures.
>
> H. J. Muller (1936a), Nobel Laureate in Genetics.

> ... it is clear that, no matter what the difficulties and the disappointments, if we want a breakthrough in human genetics we have to concentrate on methods which bypass sexual reproduction.
>
> G. Pontecorvo (1961)

> It is better for all the world if instead of waiting to execute degenerate offspring for crime, or letting them starve for their imbecility, society could prevent those who are manifestly unfit from continuing their kind ...
>
> Justice Oliver Wendell Holmes, United States Supreme Court
> Judge in the 1920s, quoted by Roslansky (1966).

Why study human genetics? When you bear in mind such quotations as those above, only a moment's thought will trigger a cascade of further questions; the structure of this double Unit hinges on the examination of many of these.

The study of human genetics should lead to an understanding of human phenotypic variation, the separation of heritable from non-heritable effects, the definition of particular gene loci and the beginnings of the definition of genetic organization in individuals as well as in populations.

What would be the *significance* and the *consequences* of discovering that part of human phenotypic variation in various characters is due to genetic differences? A knowledge of the various genotypes and allele frequencies and the patterns of marriage, ought to give us information that will help in the prediction of the range of possible types and even frequencies of particular phenotypes in a future generation.

For instance, hidden in the human gene pool (the sum of all the alleles at all loci) are some alleles that in certain situations and in particular genotypic combinations lead to the development of individuals with severely abnormal phenotypes. The presence of these alleles may have immediate, lasting and traumatic consequences for that individual and his or her family. What strategies could we adopt to minimize the occurrence of such 'abnormal' phenotypes? (The role of medical genetics in counselling, practical diagnosis, and the prevention and treatment of genetic disease is important here.)

Further, let us suppose that there *are* genetic differences within human populations. This would mean that *changes* in the frequency and type of genotypes might be occurring. This is no more or less than saying that evolution could be occurring. Let us first look to the past. There is much apparent diversity within and between different human populations. If this variation were to include some that is genetic in origin, then we could enquire about the significance of these differences. Did they originate as the result of random effects or systematic and directional effects on the gene pool? Consider almost any far-reaching event in human history—the plague in Europe, the migration of Europeans to North America, the slave trade, or the pregnancies left behind by invading armies—would such effects have affected the local gene pools significantly?

Biological evolution, you already know, depends on the existence of heritable genetic variability. Do we have any reason to suppose that such evolution has ceased in humans? Or is it still occurring? More importantly, is it proceeding in any particular direction and can the consequences of such evolution be predicted, evaluated or controlled in the same way that the genetics of many plant and animal species have been manipulated in the development of agriculture and agronomy?

Already in this Course you will have come across several examples of both the application of genetic methods to the study of human characters and the use of studies of humans to elucidate problems in genetics. So several themes that might have been expected to be covered by these Units will not be mentioned here at all.

Instead, we have chosen to:

1 describe the methods available for the study of human genetics, with particular reference to the rhesus blood-group system;

2 introduce the topic of behavioural genetics and its application to schizophrenia in humans;

3 discuss the question of human population genetics and its relationship to the idea of race;

4 give a case study of the 'race–IQ controversy';

5 consider whether the human population is still evolving;

6 consider the impact of medical genetics and its possible development as it begins to affect society.

We begin with the methods that have been developed to study human genetics.

14.1 Methods of genetic analysis in humans

Humans have few progeny and generations are long; individuals live in many different environments. Nevertheless, by the end of this Section you may think that many of these constraints on genetic analysis in humans have been circumvented by the choice of appropriate strategies—the use of pedigree studies, twin and sibship correlations, blood-protein analysis and the culture of human cells *in vitro*. Of help here is the fact that the population is large and, in general, is well scrutinized for usual and unusual characteristics, including disease.

14.1.1 Pedigree analysis

We introduced the analysis of individual pedigrees in Units 1 and 2, and in later Units you have used a similar method of deduction to solve problems. The pedigrees shown in Figures 1–3 provide an opportunity for you to make sure that you are still familiar with the analysis.

> **ITQ 1** Figure 1 shows the offspring produced from individuals affected by the condition known as brachydactyly (shortened fingers; see Units 9 and 10, Section 9.4.2). What is the genetic basis of the trait involved in the pedigree?

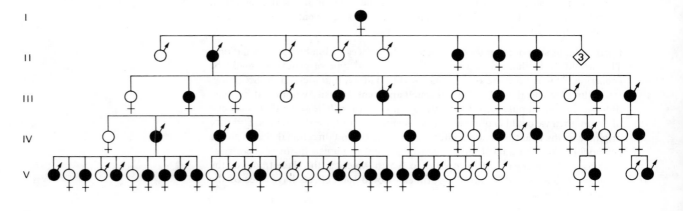

Key.

○ female

♂ male

● showing trait

◇ 3 further undetermined offspring

Figure 1 Brachydactyly pedigree (after Farabee, 1905).

The answers to the ITQs are on p. 682.

ITQ 2 What is the likely genetic basis of *phenylketonuria* (see the pedigree of Figure 2)? (The families referred to in the pedigree lived on an isolated group of small islands off the coast of Norway, so the individuals involved may have ancestors in common.)

phenylketonuria

Figure 2 PKU pedigree (after Følling, Mohr and Ruud).

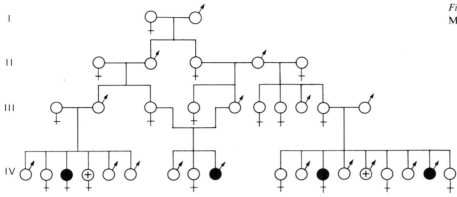

Key
⊕ probably affected but died young

ITQ 3 The affected individuals in the pedigree of Figure 3 are deaf-mutes. What genetic hypothesis can you suggest to explain the pattern of inheritance in this pedigree? Look at each marriage and sibship separately first.

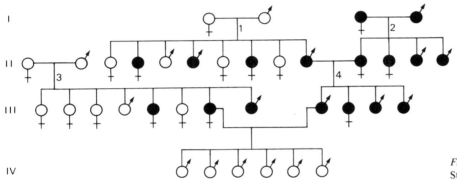

Figure 3 Deaf-mute pedigree (after Stevenson and Cheeseman).

You may already be thinking that if inherited deaf-mutism is not common (which is true) then such marriages as 2 and 4 in Figure 3 ought to be rare *by chance*. However, marriage *is* assortative for many characters (for example, height; a short person is more likely to marry another short person, a tall person another tall person), a point which we shall return to later in this Unit. So it seems quite likely that a deaf-mute will often marry another deaf-mute because they tend to be drawn together socially, and can achieve a better communication and understanding with each other than with normal people.

However, although pedigree analysis may allow us to define many human genes, recognizable by varying types of phenotypic expression, there are still difficulties in analysing this assemblage of information on pedigrees.

1 Phenotypic similarity

phenotypic similarity

If two individuals are born with similar unusual phenotypes, this does not necessarily indicate similar underlying genotypes. Indeed, any particular occurrence may only be a phenocopy (Units 9 and 10, Section 9.1.3) of a known genetic defect. There also remains the possibility that an assembly of pedigree data taken from different families and locations may turn out to be a collection of diverse genetic effects in which the pooling of data has obscured small differences in the patterns of inheritance or, for identical patterns, in the genes responsible for a given condition. This cautionary note becomes important when the recognition of the phenotype is itself a complex matter, as for instance in the diagnosis of mental disorders such as schizophrenia, or even in conditions that appear to be phenotypically quite homogeneous.

623

There is an additional difficulty that accompanies the pooling of segregation data from different sibships; this is derived from the problem of *ascertainment*. (You have already met this problem in Home Experiment 4.)

ascertainment

Let us imagine that we are attempting to determine whether a particular phenotype is due to a recessive allele. We test the observed ratio of phenotypes summed from many different families against a particular genetic expectation (for example, 1 : 1, 3 : 1, etc.). We know which sibships (that is, the children of any one couple) should be included; they will be recognized because they contain at least one individual showing the phenotype under scrutiny. Possibly you can already see how this method of collecting data for the χ^2 test would introduce a bias into the observed segregation ratio?

Let us take as an example the hypothetical gene locus *A* and its recessive allele *a*. Now consider sibships of, say, four children from a marriage in which we suspected both parents of being heterozygotes (*Aa*) because one of the children is *aa* (an 'affected' individual). Leaving aside sex, differences in these sibships will be of four types only—with one, two, three and four affected individuals). What are missing are sibships from similar marriages in which all four children are normal—because there would, of course, be no reason to abstract them from the population. This biases the ratio in favour of affected sibs, resulting in what is called *incomplete ascertainment*, because only part of the data is available for analysis.

Table 1 shows the calculation of all the possible sibships of four from a marriage between two heterozygous partners (*Aa* × *Aa*). *aa* is the recognizable homozygous recessive. Note that in the Table we have used the symbol *A*– to include the genotypes

Table 1   Probabilities and types of sibships of size 4 from *Aa* × *Aa* marriages (after Stern, 1973)

Birth order of children				Probability of sibships (see *STATS**, Section ST.3)	Number of *A*– and *aa* in 256 families		Type of sibship
first	second	third	fourth		*A*–	*aa*	
A–	*A*–	*A*–	*A*–	$\frac{3}{4} \times \frac{3}{4} \times \frac{3}{4} \times \frac{3}{4} = \frac{81}{256}$	324	0	4*A*– : 0*aa*
A– *A*– *A*– *aa*	*A*– *A*– *aa* *A*–	*A*– *aa* *A*– *A*–	*aa* *A*– *A*– *A*–	$4 \times \frac{3}{4} \times \frac{3}{4} \times \frac{3}{4} \times \frac{1}{4} = \frac{108}{256}$	324	108	3*A*– : 1*aa*
A– *A*– *aa* *A*– *aa* *aa*	*A*– *aa* *aa* *aa* *A*– *A*–	*aa* *aa* *A*– *A*– *aa* *A*–	*aa* *A*– *A*– *aa* *A*– *aa*	$6 \times \frac{3}{4} \times \frac{3}{4} \times \frac{1}{4} \times \frac{1}{4} = \frac{54}{256}$	108	108	2*A*– : 2*aa*
A– *aa* *aa* *aa*	*aa* *A*– *aa* *aa*	*aa* *aa* *A*– *aa*	*aa* *aa* *aa* *A*–	$4 \times \frac{3}{4} \times \frac{1}{4} \times \frac{1}{4} \times \frac{1}{4} = \frac{12}{256}$	12	36	1*A*– : 3*aa*
aa	*aa*	*aa*	*aa*	$\frac{1}{4} \times \frac{1}{4} \times \frac{1}{4} \times \frac{1}{4} = \frac{1}{256}$	0	4	0*A*– : 4*aa*
				Total	768	256	
				Total, omitting the 4*A* : 0*aa* sibships	444	256	

* The Open University (1976) S299 STATS *Statistics for Genetics*, The Open University Press. This text is to be studied in parallel with the Units of the Course. We refer to it by its code, *STATS*.

AA and *Aa*. The distortion in the observed ratio of *A*– to *aa* genotypes (from the expected 3*A*– : 1*aa*) is considerable and is of greater concern when sibship sizes are *smaller*. Methods for correcting this type of distortion, at least in simple cases, are discussed in the *Home Experiment Notes**. Here, we need merely note that various problems of ascertainment arise, which are often more complex than those we have treated. For instance, a particular phenotype may be the result of the interaction of two alleles at different and unlinked loci, or of a dominant allele with *partial penetrance* (that is, a dominant allele that does not produce the expected phenotype in every individual possessing that gene).

partial penetrance

*The Open University (1976) S299 HEN *Home Experiment Notes*, The Open University Press. These form part of the supplementary material for the Course.

3 Polygenic control and threshold effects

Not all data collected from sibships can be fitted by a simple genetic hypothesis, even allowing for incomplete ascertainment. Sometimes the evidence for a *heritable basis* for the phenotype is very clear from pedigrees, but the number of loci concerned may be uncertain (and therefore, not give such obvious segregation ratios), and, indeed, there may be polygenic control of the character. This situation can lead to observations that are in apparent contradiction to the basic Mendelian theory of genetics. If one examines all the collected sibships in which the first-born child showed the phenotype of *pyloric stenosis*, one finds that the probability of the occurrence of the trait in subsequent sibs *rises with the number of sibs already affected*! (Pyloric stenosis is a severe defect, in which there is a thickening of the circular muscle layer of the pylorus between stomach and intestine. It occurs in newborn infants, usually males. In the past, the condition usually led to the death of the infant but it can now be remedied by the use of relaxing drugs or a relatively simple muscle-splitting operation.)

pyloric stenosis

This observation patently will not fit the expectation of a *constant probability of re-occurrence*, that is, simple Mendelian ratios (for example, 1:2 and 1:4), in subsequent sibs and in different families, which is what genetic segregation leads us to expect.

Although one hypothesis to account for these observations would be that pyloric stenosis was entirely environmentally induced with the (unknown) environmental inducer differentially distributed among families, a genetic hypothesis would also be possible, in which there is polygenic control together with a threshold effect. Such a proposal has been advanced by Carter. He suggested that what is inherited is a *liability* to the clinical condition, and that this liability shows continuous variability in the population as the result of segregation at a number of loci (a polygenic system), but that a threshold exists above which pyloric stenosis results (Fig. 4).

polygenic control

threshold effects

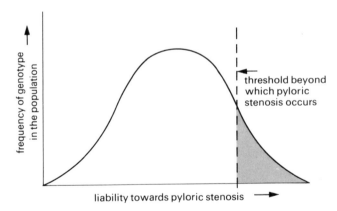

Figure 4 Liability towards pyloric stenosis.

The probability of pyloric stenosis will be the combined effect of all alleles in the polygenic system that lead to *increased* liability. If these alleles are *common* in the parents, then the genotypes of their first and subsequent children will often exceed the threshold value for expression (Fig. 5c). On the other hand, a *rare* combination showing pyloric stenosis can arise in an otherwise normal sibship (Fig. 5a). The apparent contradiction to the expectation of independent probability of occurrence of the condition in sibships is an artefact due to the lumping together of sibships from parents with very different average liabilities to the disease (that is, the pooling of heterogeneous data; see *STATS*, Section ST.4.6).

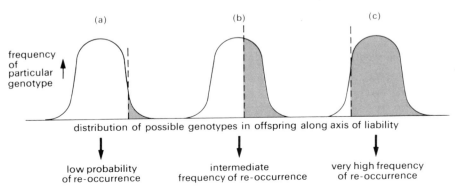

Figure 5 Different genotypes and the probability of pyloric stenosis.

The increasing probability that subsequent sibs will be affected reflects the 'weighting' given to the pooled data by sibships of parents with a high liability.

There is a second consequence of a polygenic system of inheritance: it is that the probability that pyloric stenosis will occur in related individuals will not fall off as rapidly with decreasing genetic relatedness (from sib to double first cousin, cousin, uncle, etc.) as the probability when the genetic basis is the segregation of a single gene. When we are dealing with a polygenic system with many loci, rather than with the presence or absence of single alleles, then it is likely that many of the related individuals would have sufficient relevant alleles at several of the loci to produce pyloric stenosis.

14.1.2 Twins in genetic analysis

Individuals have different genotypes, are born at different times to different mothers of varying ages and experience different post-natal environments. How in this welter of different and independent variables can the geneticist identify the genetic from the environmental components of variation in the character to be investigated? Although certain sources of variation can be eliminated by, for example, sampling only individuals of a particular age or size or sex, and the characteristic under examination can be related to other types of variation, such as age, sex or birth order, almost every human individual is likely to be genetically unique.

Contrast this situation with the common use of identical genotypes in the investigation of genetic and environmental components of variation in other experimental organisms such as grasses or *Drosophila* (Units 12 and 13). Human geneticists have seized on their only chance to match this situation by using *twins*.

Twins are pairs of sibs born at the same time to the same mother. They are of two types—either the members of the pair are genetically identical (*monozygotic*, or *MZ*, twins) because they resulted from the split of a fertilized embryo early in uterine life, or they are the result of the fertilization of two separate ova at the same time (*dizygotic*, or *DZ*, twins), and will be no more similar genetically than any two sibs not of the same age.

monozygotic (MZ) twins

dizygotic (DZ) twins

Twins have been used to investigate the relative magnitude of environmental and genetic components of variance for any chosen character in two distinct ways.

1 Monozygotic/dizygotic twin comparisons are used to estimate the genetic component of the total phenotypic variation.

2 Monozygotic twins, raised together or separately, are compared in order to examine how changes in post-natal environments affect various phenotypes.

Monozygotic and dizygotic twin comparisons

The rationale behind the method is straightforward. Table 2 shows that any difference *between members* of a MZ twin pair must be due to *environmental* effects alone, whereas for DZ twins differences in environment will interact with the existing *genetic* differences. In the simple model, it is assumed that the environment of both sorts of twins is very similar (but see p. 629).

Table 2 Components of phenotypic variation in MZ and DZ twins raised together

Twin type	Environment	Genotype	Source of phenotypic differences between members of a pair
MZ	very similar	identical	environmental
DZ	very similar (reservation mentioned later)	they have only half their genes in common (on average)	environmental and genetic

ITQ 4 Figure 6 shows the difference in height (in centimetres) between twins for 50 MZ twins, 52 DZ twins and 52 pairs of non-twin sibs of the same sex. Examine Figure 6 and then match the statements A, B and C with the conclusions (1, 2, 3 or 4) that seem most reasonable.

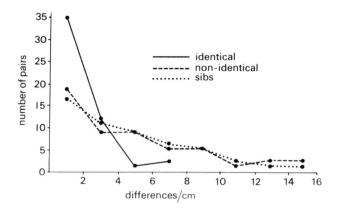

Figure 6 Difference in height between pairs for MZ and DZ twins (after Stern, 1973).

Statement

A   The difference in height within MZ twin pairs is less than that within DZ twin pairs.

B   The differences within DZ twin pairs and within sib pairs are the same.

C   The differences for pairs of sibs are greater than for MZ twins.

Conclusion

1   Being born at different times is not a significant environmental variable affecting height.

2   Environmental differences are responsible for most of the between-pair differences in height.

3   MZ and DZ twins experience very different environments.

4   There are genetic differences affecting height in this sample.

The concept of heritability was discussed at length in Unit 12. Here, we are interested in the estimate of the broad-sense heritability for certain human physical characteristics. A list of estimates is given in Table 3 (after Osborne and de George, 1959) for various characteristics; the two sexes (MZ and DZ male and female pairs) are treated differently.

Table 3   The broad-sense heritability h_B^2 of some human characteristics

Physical characteristic	MZ and DZ twin pairs	
	female	male
1 femininity–masculinity index	0.85 ± 0.06	0.78 ± 0.21
2 height (stature)	0.92 ± 0.03	0.79 ± 0.21
3 weight	0.42 ± 0.25	0.05 ± 0.94
4 cephalic index (head length : head breadth)	0.70 ± 0.13	0.90 ± 0.10
5 length of foot	0.81 ± 0.08	0.83 ± 0.16
6 length of arm	0.87 ± 0.06	0.80 ± 0.20
7 circumference of hip	0.66 ± 0.15	0.19 ± 0.93

The broad-sense heritability values for arm length, foot length and cephalic index are high and are correlated positively with stature in general, but the heritability for weight is low. It has been suggested that the difference between values for males and females for weight and for hip circumference might reflect the increased interest and success in maintaining their 'ideal' weight among the women (these twins were measured as adults). This would reduce a part of the environmental component of variance, thereby increasing the broad-sense heritability.

The same general approach using MZ–DZ comparisons is used to investigate the genetic components of threshold or continuous characters (for example, schizophrenia), but a different measure is used, that of *concordance*.

concordance

This parameter is the proportion of twins in which the character under examination is the same in the twin pair (the twins are said to be *concordant*) expressed as a

627

fraction of all pairs of twins examined (non-matching twins are called *discordant*); that is,

$$\text{concordance value} = \frac{\text{number of concordant twin pairs}}{\text{total number of twin pairs}}$$

Complete agreement gives a value of one, and no agreement would give a zero value. (Once again we refer you back to Unit 13, Section 13.9.2.)

Figure 7 indicates the concordance values found for a number of pathological situations in MZ and DZ twins.

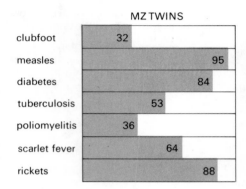
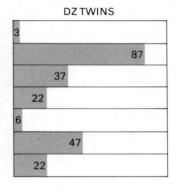

Figure 7 Percentage concordance in twin pairs affected by various pathological conditions (after Stern, 1973).

Following the same argument as before, we must say that significant differences in concordance between MZ and DZ twins (in favour of higher concordance values in MZ compared with DZ twins) indicate a genetic component in the characteristic under analysis.

ITQ 5 What does the difference in concordance values for MZ and DZ twins for poliomelitis and rickets indicate?

There is no contradiction in this conclusion (to ITQ 5). However, it does have wide implications. Even though an *obvious* non-genetic agent (like a virus or a dietary deficiency) can be singled out as the prime agent responsible for a particular phenotype, genetic differences in susceptibility (to infection and to rickets in a low vitamin D diet—possibly a threshold effect which is similar in MZ but different in DZ twins) do exist in populations and contribute to the total variation in susceptibility. We shall return to genetic variation in susceptibility to disease when examining the origins of genetic variation in human populations (Section 15.1); but, in the data of Figure 7, notice that concordance for measles is very high in both types of twin. Clearly, high concordance in both twin types does not *by itself* indicate genetic variation in susceptibility to measles, but probably rather reflects that measles is very infectious and at the age when it is commonly met the twins are likely to be together at home, passing the infection between them!

Comparisons between MZ twins reared together or separately

MZ twins have all their genes in common, so any differences (barring mutation) between members of this type of twin pair must reflect environmental differences that the two members have experienced.

In contrast to MZ/DZ twin comparisons, this situation can be used to investigate the extent to which environmental differences can affect particular phenotypes. The procedure is to identify and measure differences in a particular characteristic *between* members of a MZ twin pair, who have either been brought up together or separated and reared apart. A significant increase in the within-pair difference when the twins have grown up apart can only be explained as a response to some changed environmental factor; their genotypes remain unaltered! This method can be very informative, but is not easy to apply because MZ twins only appear at a frequency of about four per thousand births, and the estimates of those separated at birth is about one in a thousand twin pairs, so one expects to find separated MZ pairs only at a frequency of about four per million births! One also imagines that twins are separated only under extreme circumstances and it is possible that early separations are not recorded or communicated to the separated twin pair. This leaves us with the possibility that *identified* split MZ twins may be a non-random sample of all

Table 4 Data for height and weight for twins reared together and separately

| | Height | | | Weight | | |
	brought up together		brought up apart	brought up together		brought up apart
type of twin	MZ	DZ	MZ	MZ	DZ	MZ
mean difference	1.7 cm	4.4 cm	1.8 cm	1.86 kg	4.54 kg	4.49 kg
correlation between pair	+0.932	+0.645	+0.969	+0.917	+0.631	+0.886

split MZ twins. Some of the difficulties that this may produce are revealed later in the Unit in the IQ case study. Table 4 (data extracted from Cavalli-Sforza and Bodmer, 1971) is provided merely to show the type of data that can be obtained. A comparison is made with DZ twins brought up together.

Criticisms and limitations of using twins

The use of twins appears to be a powerful method for investigating the relationship between the genetic and environmental components of phenotypic variation. Precisely because evidence from such investigations has been used in support of many controversial claims, you should be clear about the reservations in the conclusions drawn from twin-studies. Below we give a number of criticisms and shortcomings of the method, together with the effect of this bias on the determination of heritability in the broad sense (Unit 13). Careful experimental design and testing can surmount some of these problems, however. Realize that the total variation is partitioned between heritable and non-heritable components and that an error in estimating the *non-heritable* fraction will, of course, affect the proportion of the variance then calculated as heritable, and vice versa.

1 The first and obvious limitation is that the correct diagnosis of MZ and DZ twins is sometimes difficult—often mere physical resemblance is used, but to be certain, other tests, for example finger and palm prints, and electrophoretic identification of blood proteins, should be carried out.

electrophoretic identity

2 The assumption that the environments of MZ and DZ twins are the same can be questioned. Even *in utero*, where members of a MZ twin pair are thought to be in an identical environment, this need not be so for DZ twins because there may be competition, for instance for nutrients, thus reinforcing the differences between the pair. After birth, it is impossible to be sure that MZ twins are treated in the same way as DZ twins by parents and teachers, whose expectations of identity in the former but not in the latter may be acted out in their interactions with the twins, whose own relationship with one another will also tend to be modified by such treatment (dressing alike, etc.). In addition, comparisons between DZ twins of opposite sexes must obviously be excluded to avoid sex differences confounding the interpretation.

3 There is a higher probability of perinatal death in MZ than in DZ twins, so not so many MZ pairs remain unbroken as DZ pairs. The attrition may be non-random and may tend to leave untouched MZ twins who are very similar (in birth weight) giving an exaggerated impression of the general similarity of MZ twins.

4 Twins may not be a representative sample of either the genotypes or the environments of the rest of the population. Twins tend in general to have lower birth weights and to develop physically and intellectually more slowly than non-twins. Unexplained anomalies, for instance, that left-handedness is three times higher in frequency among twins than among single births, where it is about 6 per cent, also occur.

5 Separated twins may be a non-random sample. It turns out that they are more likely to be from lower socio-economic classes and the reason for separation is commonly the inability of the mother to cope with two children of the same age. Nevertheless, 'separation' may merely mean boarding out one twin with a close relative, or it may mean fostering with a totally unrelated family in a different part of the country. 'Separation' is thus not a uniform treatment.

6 Unless twins for investigation can be sampled randomly from a national register, a serious ascertainment problem exists in their identification. In the past, sampling

methods have included advertising for twins in the mass media, thereby producing self-selection of, for example, twins *believing themselves to be* identical or *knowing themselves* to have been separated.

The limitations of the use of twins are summarized in Table 5.

Table 5 Problems in interpreting data from twins

Comment	Effect on estimation of heritability (in the broad sense)
(a) failure to diagnose MZ and DZ twins	poor diagnosis would identify DZ as MZ and so underestimate h_B^2
(b) variations in the environment	overestimates h_B^2
(c) selective (non-random) death before birth in MZ twins	overestimates h_B^2
(d) sampling and ascertainment problems	not known

14.1.3 Somatic-cell genetics

Explanting cells from individuals to culture medium is a standard way of obtaining cells in order to examine their karyotype, but it is also the foundation of a growing new field of genetics. This is not only because methods for detecting molecular differences in cultured cells have grown in sophistication, but also because a method has been found for *combining* genomes from different cells, including cells of different species, in a common cytoplasm. This satisfies a basic requirement for genetic analysis in somatic cells—the bringing together of genetic material from different clones (Units 9 and 10, Section 9.4.4). (In fairness, we should now reveal that it is in this context that the Pontecorvo quotation at the beginning of these Units was made.)

somatic-cell genetics

But let us consider first what scope there is for genetic analysis in cell culture. *Cell lines* can be obtained by removing specific types of cells (often from embryonic tissues, but also from mature tissues, such as fibroblasts, bone-marrow cells, etc.) to culture medium, where with suitable techniques of serial subculturing, the geneticist can establish *clones of cells*. At times and under special conditions some cells become established as lines that can usually be subcultured through many generations but the chromosome number of the cells in a line usually departs from that characterizing its source.

cell line

cell clone

To grow cells in culture requires the use of quite complex nutrient media, and small variations in the composition of the media, or the presence of biochemical inhibitors, may be deleterious to the growth of some cell lines but not others. Depending on the culture conditions, the appearance and degree of differentiation of the cells may differ markedly. Despite these difficulties, cells will proliferate under controlled conditions and can be harvested in bulk, so that cell constituents, for example, specific enzymes, can be examined. But what useful genetic information can be obtained from a particular clone about the individual from which it was derived?

> **ITQ 6** Select from the following list of phenotypes those that could possibly be studied in cultured human cells.
>
> | | | | |
> |---|---|---|---|
> | 1 | extra-nutritional requirements | 7 | haemophilia |
> | 2 | deaf-mutism | 8 | brachydactyly |
> | 3 | electrophoretic differences in proteins | 9 | height (stature) |
> | 4 | temperature-conditional lethality | 10 | hair colour |
> | 5 | pigment formation | 11 | chromosome abnormalities |
> | 6 | antigenic differences | 12 | PTC tasting ability |
> | | | 13 | antibiotic resistance |

In a moment, we shall examine some specific examples of the genetic operations that have been made possible by the use of somatic-cell genetics, but we should give a thought to how the initial cells of different genotype and phenotype are discovered, because this illustrates the 'hybrid' nature and the 'ancestry' of this area of genetics. In the first place, sampling cells from different individuals provides a possible method for discovering subtle biochemical differences (see also Section 14.2). In some instances when the genetic differences are already known (if, for example, cells are taken from galactosaemic individuals, or from individuals with known chromosomal changes), then culturing cells from individual organisms opens up the prospect of a more convenient way of investigating the detailed biochemistry of cell defects.

Another source of variants in cell culture is through their induction and selection within established cell lines. In this instance, the culture serves the human geneticist in the same way as bacterial cultures do the microbial geneticist. For example, it is possible to isolate cell lines that are resistant to 8-azaguanine from a previously sensitive clone after the induction of mutations. The obvious advantage here is that the differences between the variants can be limited, well controlled and indeed selected for in the same way as with bacteria. This contrasts with the situation when cultures of cells from different individuals are used, when there is an enormous 'background' of genetic differences between the cell lines.

There is, however, another important use that can be made of cell cultures; this depends on the technique of cell hybridization. Cell hybridization can be achieved by the fusion of somatic cells from individuals of diverse genetic make-up, or even from different species. Fusion takes place *in vitro* to form heterokaryons that are produced by simple cytoplasmic fusion, with the cell nuclei remaining separate. Hybrid cells are formed when actual nuclear fusion takes place following the cytoplasmic fusion.

The hybrid cells continue to grow and divide, and can readily be cloned. There are a variety of possible genetic uses to which the hybrids can be put, including the study of the structure of enzyme proteins and the number of genetic loci involved in enzyme activity, the regulation of growth and the control of cell differentiation and linkage. Heterokaryons, which do not multiply in culture, can be used for short-term analyses of, for example, gene–cytoplasm interaction and complementation.

In the remaining part of this Section we shall present three fairly simple genetic experiments with somatic cells to illustrate:

(a) the induction and selection of hybrid cells from two diverse cell lines (the *cross*);

(b) the use of inter-specific cell hybridization to assign a particular gene function to a single human chromosome (the process of genetic *mapping*);

(c) lastly a brief example of a *complementation* test.

The isolation of a somatic-cell hybrid by selection

Cell lines resistant to the compounds 8-azaguanine (Radio programme 2) and 5-bromodeoxyuridine lack, respectively, the activities of the enzymes hypoxanthine-guanine phosphoribosyl transferase (HGPRT⁻) and thymidine kinase (TK⁻). If these enzymes were functional the compounds that resemble the natural substrates would be taken into the nucleotide pool of the cell and would eventually kill it.

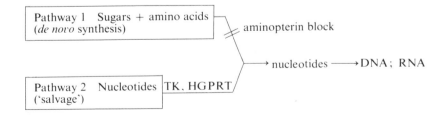

Figure 8 Pathways of nucleic acid biosynthesis.

As Figure 8 shows, the HGPRT⁻ and TK⁻ cell types have an alternative pathway for nucleic acid synthesis and therefore can still grow. This alternative pathway (pathway 1), however, can be blocked by aminopterin. Thus if TK⁻ and HGPRT⁻ cells are placed in a medium containing hypoxanthine, aminopterin and thymine (called HAT medium), both pathways are blocked, so neither cell type grows. It was found that in mixtures of the two drug-resistant cell types in HAT medium, some cells were still growing. They turned out to be *hybrid cells*, produced by fusion

hybrid cells

between TK⁻ and HGPRT⁻ cells and having the genomes of *both* initial cell types. The hybrid cells received a functional TK gene from the HGPRT⁻ cell and a functional HGPRT gene from the TK⁻ cell, as shown in Figure 9, so the salvage pathway allowed the building up of new DNA.

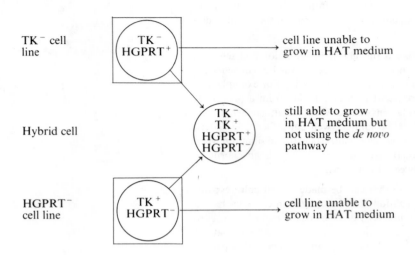

Figure 9 Enzyme properties of hybrid cell line.

There are several selective systems for recovering hybrid cells, and the fusion process can be mediated by a killed virus, but at the moment we shall only concern ourselves with the uses to which one of the *results* of cell hybridization is being put.

Mapping a gene function on a specific human chromosome

Consider the selection of hybrid cells from a mixture of mouse cells that were deficient in thymidine kinase (TK⁻; see Fig. 9) and normal human cells in a HAT culture medium. The TK⁻ mouse cells will not grow alone and the human cells grow only slowly. Any hybrid cell formed grows rapidly and can be recognized because its karyotype is the sum of the mouse and human karyotypes. Figure 10 (p. 633) shows the appearance and the karyotypes of both cell lines and of the hybrid.

After some generations of continued culturing in HAT medium, the number of human chromosomes in the hybrid cells had fallen to only three. (Human chromosome loss is a common phenomenon in cell hybrids between human and other species. The loss is often complete.) We shall call these cells type A. A part of clone A was transferred to medium supplemented with bromodeoxyuridine, which will select *against* cells showing TK activity. The surviving cells in this medium had all lost one of the specific human chromosomes of group E (ringed in the Figure) but not necessarily the other two. These cells may be called type B cells, and were unable to grow in HAT medium, indicating a gene loss associated with an E group chromosome. This process of selection is summarized in Figure 11. (Note that TK is not functional in the mouse cell line.)

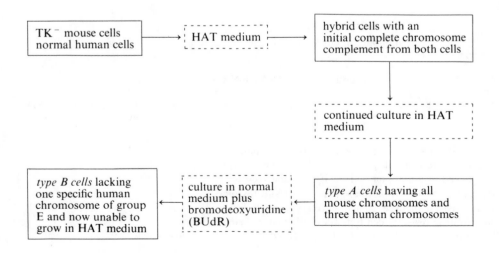

Figure 11 Selection procedures for cells of types A and B.

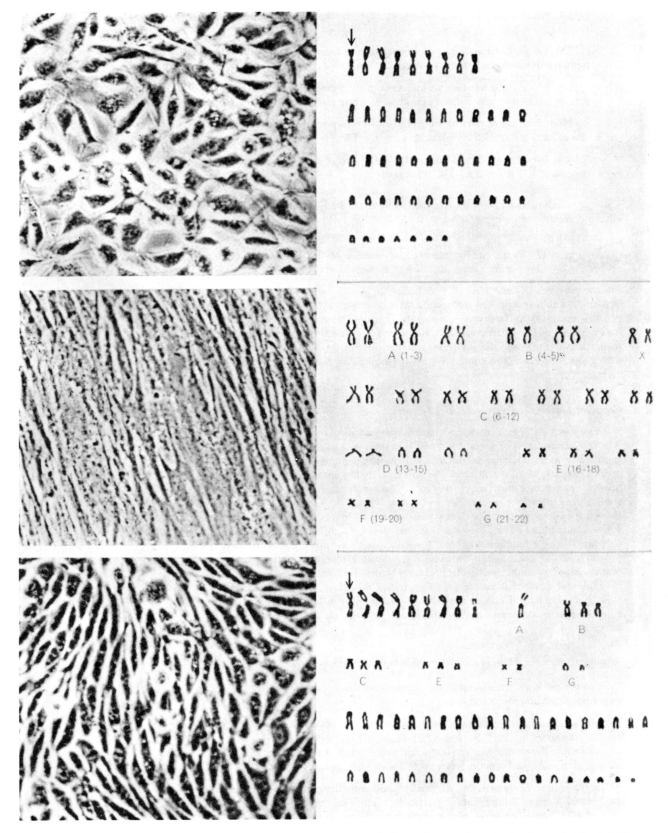

Figure 10 Mouse–human hybrids are illustrated by the cell cultures (*left*) and the karyograms (*right*) of the mouse parent line (*top*), the human parent (*middle*) and the hybrid (*bottom*). The human cells, derived from embryonic lung tissue, contain the normal number of chromosomes (46, or 23 pairs), arranged here in the usual seven groups (plus the two female sex chromo-somes). Except for a tendency to align in parallel, the hybrid cells look more like the mouse cells than the human ones. This is in keeping with the fact that the hybrid karyogram contains only 14 of the 16 human chromosomes, which are readily distinguished from mouse chromosomes.

ITQ 7 Select the correct statements (1–4) to fit the experiment just described (Fig. 11 on p. 632).

1 The TK gene of the human genome cannot be located in any of the three human chromosomes left in type-A cells.

2 One of the three human chromosomes in the type-A cells carries the TK gene of the human genome and permits this hybrid to grow in HAT medium.

3 Human chromosomes of group E, lost from type-A cells when they give rise to survivors, called type-B cells, carry the TK gene in humans.

4 The two human chromosomes left in the hybrid cell, type B, carry the genes for HGPRT and for TK in humans.

This example has shown how an enzyme function can be assigned to a particular human chromosome by using:

(a) selective media,

(b) hybrid cells differing in the presence or absence of the TK gene, and the propensity of inter-specific hybrid cells to lose *human* chromosomes.

For the purposes of these Units it is sufficient for you to understand this example in human gene mapping, but this technique is of immense value in other fields of genetics, especially in the investigation of the *control* of genetic functions. It might, for example, be that a particular 'human' gene is only functional when 'turned on' by the activities of other 'human' genes and these may be on other chromosomes. In such a situation the entire control system for expression of that enzyme could be mapped.

Complementation tests with somatic cell genetics

The first example of cell fusion did illustrate complementation, but its major use is in the *manipulation* of somatic cells by selection. As should be clear from what has already been said, inherited differences in human pedigrees in unrelated families in different parts of the world cannot be directly tested by complementation. No one is willing to accept a marriage partner merely to satisfy the curiosity of a geneticist about their genotype! However, for *some* phenotypes (ITQ 7 reminds you of which classes) it is possible to make complementation tests by fusing cells derived from *different* individuals having the *same* inherited *phenotypic* difference. Galactosaemic individuals lack a functional enzyme for galactose-1-phosphate uridyl transferase activity and therefore cannot metabolize galactose. Hence, they suffer metabolic complications when they are fed lactose from birth. There are thought to be at least two different genes that induce galactosaemia. Consequently, if there is complementation, the hybrid cells formed by fusion between cells from galactosaemic individuals from different families will be able to grow on galactose as a carbon source.

14.1.4 A comparison of the different methods used in human genetics

Each of the methods introduced in the previous Sections (14.1.1 to 14.1.3) has its own advantages and restrictions.

Table 6 summarizes the contrasts between the techniques. Note that karyotype analysis (discussed in Unit 5, Section 5.2.2) is subsumed under both pedigree analysis and somatic-cell genetics, and the limitation to the types of phenotype that can be studied by somatic-cell genetics, referred to above, still applies.

At the beginning of these Units we briefly alluded to the constraints on genetic analysis in humans (small families, lack of 'controlled' matings, complex and 'uncontrolled' environmental variables affecting the phenotypes, and, of course, the problem that the observing geneticist has the same life expectancy as the subjects). Table 6 shows that in the study of human genetics these have served not so much as constraints but more as challenges to experimental ingenuity. Additionally, do not lose sight of the fact that, although marriage, monogamy and other social patterns (for example, the family) appear to be hindrances to genetic analysis when a comparison is made with the scope of genetic manipulations in prokaryotes, fruit flies and wheat, they are a very real feature of human society and, therefore, to be relevant to the human situation at all, genetics has to come to grips with them.

Table 6   Comparison of methods in human genetics

Line of enquiry	pedigree analysis	Method twin study	somatic-cell genetics
1 evidence for a genetic component to the variation (including heritability)	yes, but confounded by environmental similarities	yes MZ/DZ comparisons	possible, tested by biochemical persistence of phenotype in cultured cells
2 identification of gene loci responsible for genetic differences	yes, e.g. PKU	no	yes, e.g. HGPRT$^-$ and differences *induced* in cell culture
3 dominance relationships	yes	yes, under certain circumstances	yes, e.g. TK$^+$ $\times$ TK$^-$
4 assigning a particular gene to a particular chromosome	yes, rarely (but see 5 and 6)	yes, under certain circumstances	yes, e.g. mapping of TK function
5 X-linkage	yes	no	yes
6 recombination values	difficult*	no	no
7 complementation tests for genetic identity of similar phenotypes	rarely and by 'chance' (cf. deaf-mutism pedigree (Fig. 3)	no	yes, e.g. galactosaemia
8 analysis of biochemical phenotypic differences	yes, e.g. blood-group proteins	no	yes, e.g. galactosaemia HGPRT$^-$
9 analysis of complex phenotypes (morphological and behavioural)	yes, e.g. deaf-mutism and pyloric stenosis	yes, in principle, e.g. susceptibility to disease	no
10 extent and effect of environmental variation on the phenotype	no, not controlled except by statistical methods, e.g. correlations with age of mother, number of children, etc.	yes, in principle, comparison of MZ twin pairs reared apart and together	yes, but a narrow range of conditions, e.g. media and temperatures on cellular phenotypes

* It is important to recognize that a *correlation* between the appearance of two phenotypes is no evidence of linkage. Instead, one has to search out marriages of the relevant double heterozygote, and for recessive alleles, for example a, b, one needs a marriage between such a person $a/ /a^+ ; b/ /b^+$, and the double homozygous recessive individual $a/ /b$; $b/ /b$ (see *STATS*, Section ST.4.7, contingency χ^2).

14.2   The application of genetics to human disease: the rhesus blood-group system

With the background knowledge of the methods of human genetics discussed in the previous Section, we may now consider in more detail a specific example, the rhesus blood-group system (first discussed in Unit 1, Section 1.8.2), the understanding of which has resulted from the application of genetic methods to the solution of complex problems that had defied interpretation by other means. We begin by looking at the history of the discovery of the rhesus system.

genetic disease

In 1939 Levine and Stetson had implicated an unknown fetal blood-group antibody in the haemolytic disease affecting newborn infants known as erythroblastosis fetalis. This disease was a common cause of infant mortality; for example, during the years 1948–1957 there were 1 991 cases in north-east England, that is, approximately five per thousand births. Before effective treatment of infants began in 1947, almost all of these babies died before they reached the age of one month.

At about the same time (1940) Landsteiner and Wiener discovered a new blood-group system as a result of injecting red cells of the rhesus monkey (*Macacus mulatta*) into rabbits and guinea-pigs and thereby obtaining an antibody that agglutinated (Units 9 and 10, Section 9) the red cells of 85 per cent of a sample of (human) blood donors in New York. Those people who possessed an antigen that reacted with the anti-rhesus antibody were called rhesus positive. It took only a short while to match up the anti-rhesus antibody with the antibody discovered earlier by Levine and Stetson and subsequently to show that erythroblastosis fetalis was the result of rhesus blood-group incompatibility between mother and child.

Since that time the management of haemolytic disease of the newborn and the working out of the blood-group system have advanced in parallel.

14.2.1 The treatment of haemolytic disease

At its simplest, Rh positive can be considered as being determined by the dominant allele Rh^+, whereas Rh-negative individuals are homozygotes, $RhRh$. The genetics will be elaborated in Section 14.2.2. The story of *haemolytic disease* of the newborn caused by the rhesus blood-group system is very intriguing, particularly because it has been of personal importance to a large number of people. It should be understood that symptoms of haemolytic disease can be caused by blood-group systems other than rhesus, but in what follows the rhesus blood-group system will be considered essentially in isolation.

haemolytic disease

The disease occurs, apparently, only when a Rh-negative mother carries a Rh-positive child. Under these conditions the mother *may* develop an anti-Rh-positive antibody (the so-called anti-D), which crosses the placenta, enters the fetal blood and begins to destroy the infant's red cells.

How often would we expect this to happen? Table 9 (p. 639) shows that the relevant (*cde*) chromosome occurs at a frequency of 39 per cent and this corresponds with a genotype frequency of $(39/100)^2$ (Units 9 and 10, Section 9.4.2), that is, 15.2 per cent. If mating in humans were random, we would expect 9.3 per cent of all babies born to be the Rh-positive children of Rh-negative mothers. Another way of looking at this is that 61 per cent of all births to Rh-negative mothers should be of children with the disease. This does not happen! Less than 5 per cent of Rh-positive children born to Rh-negative mothers actually develop the disease; that is, only 5 per cent of Rh-negative mothers carrying Rh-positive children develop anti-D. Why is this?

rhesus positive, rhesus negative

In the first place, it is very unusual for a mother to develop anti-D while she is carrying her first Rh-positive child. Normally, the fetal and maternal blood supply to the placenta are kept apart by two layers of cells and the red cells cannot cross this barrier. At some stage during pregnancy or at the time of delivery, some Rh-positive fetal red cells gain access to the maternal blood circulatory system. The cells of the mother that produce antibody *may* then be stimulated to produce anti-D (see Fig. 12). So, clearly, the formation of anti-D depends on the fetal red cells arriving in the maternal blood, and as this is a fortuitous event it cannot be predicted. Furthermore, as you will see shortly, the response of the mother is also variable and so adds another chance element to the prediction.

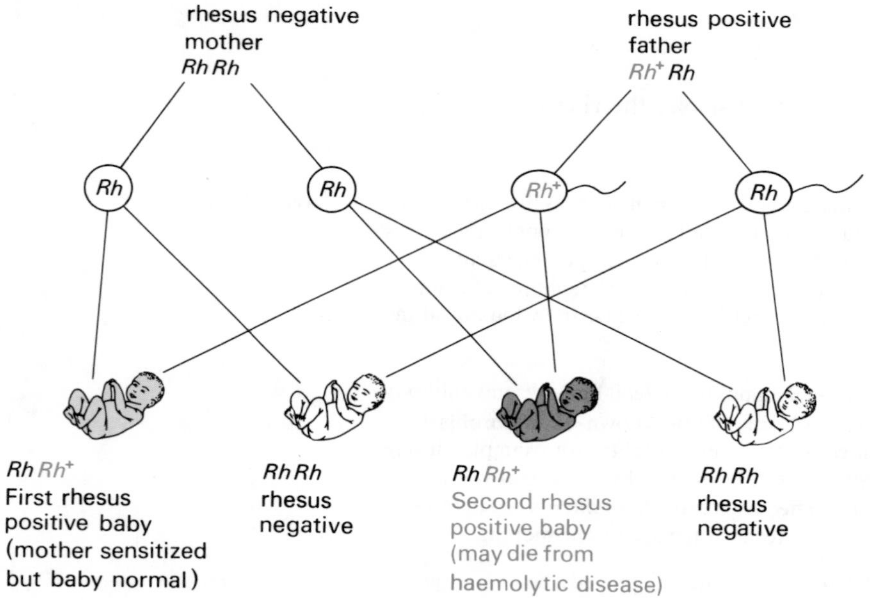

Figure 12 The rhesus system and pregnancy.

The original treatment methods were 'exchange transfusion' whereby 180 cm³ of whole blood per kilogram of body weight of the infant was transferred into the infant via the umbilical cord. This was not always successful and could not, of course, be used on the 16 per cent of diseased children who were born dead. Prevention could be the only final answer.

The clue as to how this could be achieved came from epidemiological studies*, which showed some interactions with the ABO blood-group system. It was noticed

---

* Epidemiology is the study of the distribution of disease across population and time.

that mothers who were O, Rh-negative very rarely, if ever, produced anti-D after bearing an A, Rh-positive infant. This is because the plasma of the mother contains anti-A (Units 9 and 10, Section 9.3). Perhaps this antibody destroys the invading fetal A cells *before* the Rh-positive antigen can induce the production of anti-D by the mother. Let us assume that this interpretation is correct; we can then argue that, if it were possible to remove all stray Rh-positive red cells before they can induce anti-D production by the mother, a means of preventing the disease would have been discovered. Can we inject into the mother, at the time of highest risk, an antibody that will react with the alien Rh-positive cells *before* they can induce anti-D production? Two research teams, one in Liverpool and the other in New York, found the answer at the same time, although independently.

The question was posed in this way. Knowing that the injection of blood group O, Rh-positive cells into a person of blood group O, Rh-negative stimulates anti-D production, does the simultaneous injection of anti-D with the O, Rh-positive cells reduce or prevent anti-D formation?

After experimenting by injecting Rh-positive red cells together with a dose of anti-D into volunteer males, it was eventually discovered that an 'incomplete' anti-D is most effective at preventing the development of new anti-D. This incomplete antibody coats the antigen (without causing agglutination) so that it does not make contact with the cells that form antibodies. Being satisfied with the results of their studies using volunteers, the Liverpool group gave the anti-rhesus immunoglobulin to 131 selected first-baby mothers within 48 hours after delivery. In another sample, 139 selected mothers were untreated and these acted as controls. By selected is meant that the mother was Rh-negative, the child Rh-positive and they were compatible on the basis of their ABO blood group. Six months later all 267 mothers were tested for the presence of anti-D in their sera. The results are given in Table 7.

Table 7   Effects of treatment of Rh mothers

	Anti-D produced	Anti-D not produced
controls	29	110
treated	1	130

It was clear that the administration of the anti-rhesus immunoglobulin was having the desired effect. The technique is now used extensively to prevent haemolytic disease of the newborn, although the evidence quoted above shows that it is not 100 per cent effective. The failure rate in Finland has been estimated at approximately 0.35 per cent (1975).

Without carrying the story of the clinical treatment of the disease further, we may ask the question that must interest the geneticist: Why, in population genetic terms, does haemolytic disease occur with the frequency that it does? That is, what could be the selective advantage of the rhesus-negative character? In the past, approximately 4 per cent of all children born to Rh-negative mothers suffered from rhesus haemolytic disease and did not survive infancy. Thus Rh-negative women have a lower reproductive fitness than their Rh-positive sisters. In spite of this the Rh-negative character remains in European populations (we discuss its distribution between human populations in Section 15.2.5) at a frequency of 15 per cent; that is, the *Rh* allele constitutes approximately 40 per cent of all the rhesus alleles!

QUESTION Why does the frequency of *Rh* not decline?

ANSWER One possibility might be that parents overcompensate for the death of their newly born children by having families larger than average.

It has been shown algebraically that even a little overcompensation could actually increase the frequency of *Rh*. Is there any evidence for reproductive compensation as far as rhesus haemolytic disease is concerned? Several studies have given no evidence for compensation at all, and the one study that does show compensation is suspect because it is based on a highly selected sample of nearly 11 000 mothers, all of whom had borne at least two children previously. The best designed study followed up the reproductive behaviour of 52 couples who had had a child that had died from haemolytic disease (usually due to *Rh*). The mean age of the mothers was 29.12 years. From the data available for the whole of the area in which these couples lived, it was estimated that 52 women with a mean age of 29.12 years would be expected to produce 22.07 children during the following four years. In the group under study, there were 22 such births altogether! Admittedly this was a very small sample, but it is scarcely evidence of compensation. So we are left without a clear answer.

14.2.2 The genetics of rhesus

From the purely academic point of view, the rhesus blood-group system is splendidly complicated. Fortunately those aspects that are of clinical importance are much simpler. Wiener regards all the different antigenic types as being produced by alternative alleles at a single locus. On the other hand, Fisher suggested in 1943— long before there was any clear understanding of the structure and function of genes at the chemical level—that there were at least *three* closely linked genes involved; that is, there was a supergene complex of at least three loci. Fisher spotted that the pattern in the reaction of the seven different red-cell types to the four antibodies available was of *genetic* significance.

The data Fisher used (derived from the work of Race) are given in Table 8 (after Clarke, 1968). You can see that the reactions in the first row and the fourth row are antithetical; that is, where one entry is + (agglutinated) the corresponding entry with the other antibody is − (not agglutinated) and vice versa. Consequently, these antibodies were given names that show their relationship—hence anti-C and anti-c. Fisher supposed that the antigens recognized by these two antibodies were produced by different alleles of one gene. He then predicted antithetical antibodies to anti-D (called anti-d) and anti-E (called anti-e) and a further antigenic type CdE; for the missing CDE anti-C and the CDE anti-D he predicted +, +.

Table 8   Rhesus antigens and antibodies

| Antibodies | Red-cell antigen phenotypes | | | | | | |
	CDe	cDE	cde	cDe	cdE	Cde	CDE
anti-C	+	−	−	−	−	+	?
anti-D*	+	+	−	+	−	−	?
anti-E	−	+	−	−	+	−	+
anti-c	−	+	+	+	+	−	−

+ agglutinated   ? not tested
− not agglutinated   * the original anti-rhesus antibody

ITQ 8 Expand Table 8 to include Fisher's predictions for antigens CDE, CdE and antibodies anti-d and anti-e.

Look at Table 8 again. Fisher noticed that the cde antigen was not agglutinated by the first three antibodies. Because the rhesus-negative *phenotype* had been shown to be recessive (*RhRh*), he deduced that the anti-C, anti-D and anti-E antibodies were detecting 'dominant' antigens.

He therefore ascribed symbols to these antibodies indicating that they agglutinated antigens produced by the 'dominant' alleles of the rhesus genes. As we have seen, he called these antibodies anti-C, anti-D and anti-E. Consequently, we can regard the anti-C as detecting the antigen produced by the *C* allele, anti-D detecting the product of the *D* allele, etc. However, this idea of dominance is satisfactory only for the establishment of the terminology, because with the full set of six antibodies it is clear that all the alleles involved are autonomous (codominant).

codominant alleles

Although it appears that an anti-D antibody does not exist, all the other predictions made by Fisher have been confirmed. The study of pedigrees shows that the supergene complexes are inherited in a straightforward way and families can be analysed in terms of *C/c*, *D/d*, *E/e*, as has already been implied. There are eight basic supergene complexes and, although it is again prejudging some of the evidence, it makes the explanations simpler if we regard the complexes in terms of chromosomes (see Table 9; after Stern, 1973).

Table 9   Frequencies of rhesus phenotypes and chromosomes

Red-cell phenotype	Chromosome	Frequency in England %
CDe	CDe	41.0
cDE	cDE	14.0
cde	cde	39.0
cDe	cDe	2.5
cdE	cdE	1.2
Cde	Cde	1.0
CDE	CDE	0.2
CdE	CdE	very rare

Using the four antibodies given in Table 9, it is possible to distinguish people heterozygous at the C locus, but not at the D or the E loci. This means that we have phenotypic classes that include both homozygotes and heterozygotes. For example, some data for the United Kingdom are given in the first two columns of Table 10 (Race et al., 1948). As you can see from the total, the other possible phenotypes must have been very much more rare.

Table 10   Frequencies of rhesus genotypes

Observed genotype	Frequency/%	Estimated genotype
$\dfrac{CDe}{C?e}$	16.88	$\dfrac{CDe}{CDe}$
$\dfrac{CD?}{c?E}$	12.99	$\dfrac{CDe}{cDE}$
$\dfrac{CDe}{c?e}$	35.71	$\dfrac{CDe}{cde}$
$\dfrac{cde}{cde}$	13.64	$\dfrac{cde}{cde}$
$\dfrac{cDE}{c??}$	18.18	$\dfrac{cDE}{cDE}$ or $\dfrac{cDE}{cde}$
	Total 97.40	

The only genotypes that can be distinguished without ambiguity are $cde//cde$ and $Cde//Cde$. Because $cde//cde$ is so common (13.64 per cent in this sample), and by assuming that the genotypes occur in binomial proportions, we can make some shrewd guesses about the true genotypes of the people tested. The basic criterion is that if an individual contains one c and is homozygous, ee, it is very likely that one of the chromosomes is cde. That allows us to write $CDe//c?e$ as $CDe//cde$, and we can deduce that this is the commonest class of all. Like cde, CDe must also be a common chromosome. It is not difficult to show that cDE is also common. We can, therefore, write down the estimated genotypes and these are listed on the right of Table 10.

Using his method of 'maximum likelihood' Fisher estimated the frequencies of the various chromosomes in the English population and these were given in Table 9.

These data convinced Fisher that there were at least three loci involved. He noticed that the frequencies of the various chromosome types in England bore a striking resemblance to the results of linkage backcross experiments (Table 9). Thus, for example, by crossing over between the heterozygous loci in the three commonest heterozygous genotypes

$$\frac{CDe}{cDE} \quad \text{gives} \quad CDE \text{ and } cDe$$

$$\frac{CDe}{cde} \quad \text{gives} \quad Cde \text{ and } cDe$$

$$\text{and} \quad \frac{cDE}{cde} \quad \text{gives} \quad cdE \text{ and } cDe$$

All three produce *cDe*, and so one can predict that the frequency of *cDe* should be the sum of the frequencies of *cdE*, *Cde* and *CDE* (1.2 + 1.0 + 0.2 = 2.4)—which it is. Although the genes had been ascribed an arbitrary order—*C*, *D* and *E*—these data can also be used to determine the order of the genes.

ITQ 9 Look at the frequencies of the cdE, Cde and CDE antigens and predict the order of the loci.

This can be visualized by using a cube in which opposite faces represent the allele pairs *D/d*, *C/c* and *E/e* (Fig. 13).

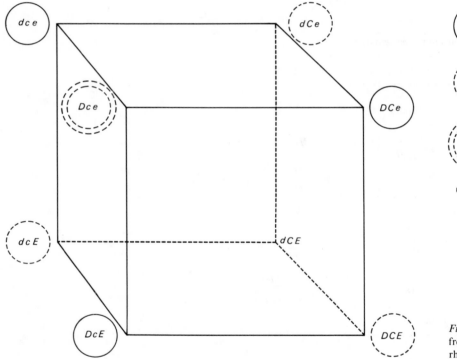

○ = three commonest chromosomes

◌ = three chromosome types produced at the same time as *Dce*

⊚ = commonest recombinant chromosome

uncircled = rarest chromosome

Figure 13 Model to explain relative frequencies of the eight genotypes of the rhesus blood-group system.

The front face represents *D* and the back face *d*, the bottom face represents *E* and the top *e*, etc. There are eight corners and these can represent the eight genotypes, because each corner is the only place at which three faces are in contact. There is only *one* way by which this Figure can be labelled, although, of course, it can be drawn with any one of the six faces at the front. When we now form heterozygotes by taking genotypes at the opposite corners of any square, it is possible to obtain the genotypes at the other two corners by a single cross-over between the relevant markers. Thus *DCe* and *DcE* on the front faces will give, as we have already seen *DCE* and *Dce*. Now, see whether you can use the model.

ITQ 10 Why is the *dCE* antigen so rare?

The model does, therefore, appear to fit the facts.

Since Fisher suggested this explanation of the relationship between the genes determining the rhesus blood-group system, many complications and clarifications have been discovered. For example, numerous alleles of D, C and E have been found, of which C^w and D^u are by far the most common; interactions occur between the D and C and between the C and the E loci, but *not* between D and E; and an interrelated group of people with the genotype $D--/ /D--$ have been studied.

> **ITQ 11** What are the ways in which two of these complications lend credence to the existence of a supergene of three gene loci with the order DCE?

Although much of the data quoted suggests that Fisher's model fits the facts, it is still fair to say that it is speculative. Certainly, some of the antigens behave as if they are controlled by single alleles, for example, C, c and C^w, and E and e. It is the crossing-over idea that is the problem. The original English data looked like the results of a linkage backcross. Data from other parts of the world have no similar pattern.

Furthermore, if Fisher is correct, it should be possible to find people in whom at least one rhesus chromosome arose as a result of a cross-over in one or other parent. This has proved an almost hopeless task. Mutation, suppression and legitimacy have all to be considered and, when you remember that only certain parental combinations of genotypes can be distinguished unambiguously, the chance of finding a suitable family is very small. Because Fisher's scheme does not fit all the facts, the notation of Wiener is still used, particularly in the USA. It has not been introduced in this Unit because using both systems at once could almost inevitably lead to confusion. Fundamentally, the difference between the two systems is that Fisher's is a genotypic one, whereas Wiener's concentrates on the phenotype. The model has had the virtues of being relatively simple and possible to test.

14.3 Behavioural genetics

We now wish to change themes. Before long, we shall introduce a discussion of the very complex topic of the application of genetics to the study and analysis of human behaviour. However, we cannot begin to do this without first considering the general question of what is meant by 'behaviour' in the sense in which geneticists can study it. Up till now in this Course we have had no occasion to raise this question (though TV programme 8 did ask some questions about the behavioural genetics of *Drosophila*). So, in this Section, we begin by discussing the concept of behaviour as a phenotype, and then turn to behavioural genetics in mammals, to provide a reasonably sound basis on which we can build our discussion of the genetics of human behaviour. Hence, the next two Sections, 14.3.1 and 14.3.2, represent something of a 'non-human' interlude.

behavioural genetics

14.3.1 Behaviour as a phenotype

We have defined the phenotype as the expression of the genotype of an individual in conjunction with both the direct effects of the environment and of interactions *between* the genotype and environment. However, you will have noticed throughout the Course that we have used the concept of the phenotype to mean a very large number of different things, at different levels of analysis. Thus, at one level we have talked about properties of the whole organism as phenotypes, for example wing shape or eye colour in *Drosophila*; at another level we have referred (Unit 5) to the chromosome itself as a phenotypic expression of a particular genotype; at yet another level we have talked about the presence of particular enzymes, or particular amino acid sequences in a given protein, as examples of the phenotypic expression of genes. All these, of course, are proper uses of the concept of the phenotype, but the result of using the term in these different contexts is often a bit confusing, especially when we attempt to analyse what is meant by the environment with which the genotype interacts in the expression of the phenotype.

behavioural phenotype

> QUESTION To make this clear, think about what is meant by the *environment* in relation to the expression of an individual enzyme such as β-galactosidase in bacteria.

ANSWER The *environment* here refers to the presence or absence within the cell of a particular class of substances—the inducer (Unit 6, Section 6.6). Hence the 'environment' is internal, being the physical and chemical milieu immediately surrounding the bacterial chromosome.

environment

Contrast this with the concept of the environment we mean when we say that certain *Drosophila* mutants are lethal or express an abnormal phenotype at temperatures above 28 °C. Here, the environment is external to the organism as a whole: we are talking about the phenotypic expression of a mutation that affects the whole organism in a particular environment. At first sight, these two examples appear different, but, when one remembers that the inducer of β-galactosidase biosynthesis has to be administered outside the bacterial cell and that we have said nothing about the biochemical reasons for the sensitivity of this lethal mutation in *Drosophila*, it becomes clear that it may not be easy to distinguish between the two uses of the word environment—the internal and the external environment of the organism (or indeed of the cell). Further complications are introduced when we consider that among contributions to the internal environment in which any given gene expresses itself are the phenotypic expressions of the entire genotype of the organism *except* the particular alleles at the locus we are considering.

Just as the expression of a given gene interacts with that of all others in the production and modification of the internal environment of the organism, so the individual organism does not merely passively respond to its own external environment but reacts to it and may modify it as well. When a phenotype is changed because of a modification of the external environment, that change may in its turn affect the environment (for instance, with organisms growing in culture media, a change in the medium may induce particular enzymes, altering the metabolism and excreted metabolic products, which in their turn affect the environment of the organism— and also of other adjacent organisms). The concept of 'environment' is thus a dynamic, interactive one. For any organism, it includes, for example, other organisms present around it. What is more, the effects of a given environment are not constant throughout an organism's life cycle, but, as we saw in Unit 8, differ during different phases of development. A slight change in the environment during one period of an organism's development may have profound consequences; at other times the same change may have negligible consequences (the effect of a drug such as thalidomide during fetal development is an obvious example here).

The point of re-emphasizing the complexities inherent in the concepts of phenotype and of environment, and the need to consider them as dynamic, interactive processes rather than static 'objects', is that these difficulties are important for all continuously varying characters and must be of major importance for behaviour as a phenotype for genetic analysis.

Consider, for example, the behavioural properties of some bacterial species; if a drop of a concentrated solution of glucose is placed in a liquid medium, many bacterial species will migrate through the medium towards it (positive *chemotaxis*). Such migration is made possible not merely by the possession of cilia which, by their beating, propel the organism, but by the presence in the cell membrane of proteins (*glucoreceptors**) that can detect and respond to the presence of glucose molecules.

chemotaxis

glucoreceptor

ITQ 12 Identify three types of mutant that might fail to respond to a high concentration of glucose in the medium by migrating towards it.

However, with more complex multicellular organisms, behavioural geneticists are inclined to be more selective in their interests. In *Drosophila*, mutations affecting the wing shape presumably affect the capacity to fly and hence behaviour. Eyeless mutants will presumably have modified behaviour. But behavioural geneticists are really concerned with the ways in which genetically produced modifications in the organism's nervous system may be reflected in definite changes in behavioural response and the use of these changes as tools for understanding the relationship of the nervous system to behaviour.

behavioural mutant

In recent years such studies have begun to make considerable progress, particularly in *Drosophila*. Hirsch and his colleagues at Urbana, Illinois, have pioneered

* These interact with the glucose molecules rather as an enzyme with its substrate.

techniques for the selection of *D. melanogaster* lines based on their response to gravity, using a mass screening maze that sorts flies as they migrate towards food according to whether they tend to climb upwards (*negative geotaxis*) or downwards (*positive geotaxis*). Hirsch and his colleagues followed this by chromosomal analysis and mapping in the offspring. This work has been followed up both by Hirsch's group and by Benzer using light-sensitive mutants (see the offprint associated with Unit 8, *The Genetic Dissection of Behaviour* by S. Benzer). In a similarly conceived and ambitious research programme at Cambridge, Brenner, working with a small nematode (*Caenorhabditis elegans*), has been able to isolate a number of behavioural mutants, generally on the basis of characteristically abnormal movements, which can be subjected to genetic analysis. At the same time, working on the hypothesis that the impaired movement may result from an altered structure of the nervous system, Brenner is attempting the complex task of analysis of the 'wiring diagram' of the nervous system in both the wild-type and mutant forms. It is too early yet to be sure how successful such an approach will be.

geotaxis

14.3.2 Behavioural genetics in mammals

When we turn from bacteria or *Drosophila* to the genetic analysis of behaviour in mammals, we immediately confront a range of obstacles that are much greater than those outlined in the previous Section.

QUESTION What are the two major classes of such obstacles?

ANSWER 1 The longer generation time of mammals, which in general makes genetic analysis slow.

2 The much greater complexity of mammalian behaviour and the difficulty of quantifying it in terms appropriate for analysis.

The study of certain mammalian systems has begun to yield information, though. For instance, there is a group of related recessive mutations in the mouse in which the homozygotes develop normally for the first week or two after birth and then begin to show increasingly severe motor defects giving rise to characteristically different types of abnormal movement; the mutations are called, descriptively, 'jimpy', 'weaver', 'reeler', and so forth. The disorders rapidly become more severe and the animals die. The behavioural defects seem to result from a developmental failure in which certain groups of cells, mainly in the cerebellum—a region of the brain concerned with the control of movement—do not migrate to their appropriate sites or make the right nervous connections with neighbouring cells during development. These then, can readily be seen as phenotypic developmental failures resulting from particular genotypes.

Can we go further, and examine the genetics of behavioural patterns that are more complex than motor defects, which are relatively straightforward? There are indeed some suggestions that this is possible. For example, a few years ago it was observed that often, if a mouse is introduced into a cage containing a member of one strain of laboratory rat, the rat will kill the mouse. Rats of other strains, however, will ignore the intruder. The selection of strains and crossing studies have suggested that this mouse-killing behaviour, in the environment in which it occurs, is under a measure of genetic control. However, modifying the environment, for instance by rearing the *killer rats* from infancy in the presence of mice, or cross-fostering the offspring of killer and non-killer rats, will also modify the behaviour.

killer rats

Similarly, intensive efforts have been made to select for a general behavioural trait such as *learning ability*. There are now well-established strains of *maze-bright* and *maze-dull* rats that perform differently when trained to run a maze for a food reward or to escape punishment. The rationale of such experiments is, first, to select strains that perform differently in a particular test situation, by breeding for the trait through several generations, as has been described in earlier Units (for example, Unit 13). Next, the extent of the genetic control of the difference in performance is examined by means of crossing studies. Finally, the possible 'intervening variables' between the gene and its behavioural phenotype are explored by, for example, trying to find characteristic enzyme differences between the strains in particular regions of the brain. However, it is here that the complexities begin to emerge. Rarely does it appear

learning ability
maze-bright, maze-dull

that the differences selected for are single-gene effects. Also, if an enzyme difference is found between, say, mouse-killing and non-mouse-killing rats, it is almost impossible to be sure that such a difference is necessarily and exclusively related to the behavioural difference observed, rather than a side effect of other aspects of the differences between the two types of animal.

Furthermore, the design of these experiments is very important. It is no use using rats from one strain and making a comparison with those of another because we have no means of telling whether any differences in behaviour between the strains result from innate differences among the individuals or in the teaching abilities of their mothers, which may or may not be genetically determined. Thus, genetic analysis must involve reciprocal crosses between the strains to determine whether or not there are any maternal effects. In addition, undetected maternal effects can be assimilated into the design of the experiment by randomizing all the youngsters born on a given day among all the mothers who gave birth on that day.

To appreciate more fully the relevance of these problems, consider the following hypothetical experiment. Rats are tested in a maze in which they have to learn to turn right in a T-maze to escape from a mild electric shock, which they receive through their footpads. Rigorous selection and breeding of two groups—those that learn the maze in the smallest number of trials, and those that learn it in the greatest—through ten generations produces two lines, one of which (A) takes an average of five trials to learn the maze, the other (B) requires an average of twenty trials. Analysis of the brains of animals of the two strains in the tenth generation shows a high level of the transmitter enzyme acetylcholinesterase in a particular region of the brain, the frontal cerebral cortex, which is associated with motor behaviour, in strain A. Both of the reciprocal crosses between the two lines give offspring that take an average of ten trials to learn the maze and have intermediate levels of acetylcholinesterase.

ITQ 13 Which of the following possible interpretations of the experiment is valid?

(i)   Strain A is a faster learning strain than strain B.

(ii)   An elevated acetylcholinesterase level in the frontal cerebral cortex causes faster learning of a maze.

(iii)   The differences between strains in learning ability probably have a genetic component.

(iv)   There are probably genetic components in the differences between the strains, but these are not necessarily associated with learning.

(v)   None of the above.

ITQ 14 How might interpretation (i) in ITQ 13 be tested further?

In fact, when experiments of the type described in the answers to ITQs 13 and 14 are performed using real strains of maze-bright and maze-dull rats, it is generally found that the differences between them in 'learning ability' disappear, or are even reversed; that is, the different strains have different learning abilities for different tasks. The point is that a very large number of factors go to make up such a complex behavioural pattern as learning, including, for instance, perception (how well the organism perceives the signal to be learned), attention, emotional state, and so forth. Although these effects are included in the phenotype of the learning of a particular skill in any *particular* situation, they cannot readily be generalized to such broad categories as a loosely defined 'learning ability'.

All this is further complicated by phenotypic plasticity. Changing the environment may change an animal's capacity to learn as a consequence of its developmental experience. Recent experiments with strains of 'dull' and 'bright' rats have shown that differences between them in the rates at which they learn particular tasks may occur in adults reared from infancy in 'restricted environments'—a term used to describe plain, isolated cages and no opportunity for social interaction or handling. On the other hand, if the young are reared in complex, communal environments with many 'toys' and much handling (conditions that are known to result in changes in brain structure and behaviour) the differences between the strains tend to disappear.

14.3.3 The genetics of human behaviour

All that we have said up to now about the problems of genetic interpretations of behavioural differences in other mammals applies even more strongly to human behaviour, where the possibilities of experimental selection and manipulation of traits must be replaced by observational methods combined with pedigree analysis, with all the attendant ascertainment problems. None the less, as a result of folk observation, there has been a long history of belief in the inheritance of 'temperament'; later, there were attempts to quantify it. Moroseness, madness, genius, business and sporting abilities, musical talent . . . , the list of characteristics that popular observation would claim 'run in the family' is very long. Of course, for every case a counter-case could be cited, and attempts to put the genetic study of human behaviour on to a sound scientific footing are faced with two fundamental problems that must be resolved before any estimate of a possible genetic, or inherited, contribution to differences among individuals can be made.

QUESTION What are these two main methodological problems?

ANSWER 1 The problem of defining and measuring behaviour according to some sort of scale so that individuals can be compared.

2 The problem of resolving the genetic and environmental contributions to any behavioural phenotype in an organism for which environmental effects cannot be controlled or randomized.

Son may indeed resemble father and daughter resemble mother, but when son and daughter have not merely been born but have also been brought up by father and mother, the parents have contributed *both* the genotype *and* the environment to the developing children.

Hence, for a very long period, even attempts to study the genetics of behaviour 'scientifically' were at the level of anecdote. Thus, we find Charles Darwin, who was very interested in the question, quoting

> A gentleman of considerable position was found by his wife to have the curious trick, when he lay fast asleep on his back in bed, of raising his right arm slowly in front of his face, up to his forehead, and then dropping it with a jerk so that the wrist fell heavily on the bridge of his nose . . . Many years after his death, his son married a lady who had never heard of the family incident. She, however, observed precisely the same peculiarity in her husband, but his nose, from not being particularly prominent, has never as yet suffered from the blows . . . One of his children, a girl, has inherited the same trick . . .

But the inheritance, or otherwise, of such foibles, is of relatively little interest. In genetic studies, from Darwin's day to the present, most of the attention of human behaviour has been concerned with two major themes: the inheritance of intelligence and the inheritance of mental illness. The history of these interests has also been deeply intertwined, as we have seen in the history text*, with ideas on eugenics and of individual, racial and class superiority. In these Units we are going to consider only one example in detail, that of the inheritance of intelligence, or, more accurately, of the capacity to score within a given range on the intelligence quotient (IQ) scale. But before we do so, we can briefly review some of those few cases in which there is known to be a relatively simple relationship between an altered genotype and a behavioural pattern.

14.3.4 Single-gene effects and mental retardation

Some 2 per cent of children in Britain are regarded as mentally retarded; that is, when brought up in a 'normal' environment they fail to learn normally; in severe cases, they may never be able to speak, or coordinate their movements beyond those of a baby or a two-year-old child; in others the deficit may be much slighter and associated with other 'behavioural problems'. In a small proportion of cases the retardation is clearly associated with obvious brain damage, occurring either before or after birth. In most, however, there are no obvious signs of brain damage. In a very small percentage of cases (some 2–10 per cent of all mentally retarded), the retardation can be shown to have a clear genetic association. The best known of such genetic diseases is phenylketonuria (PKU; see Section 14.1.1), which has been

mental retardation

* The Open University (1976) S299 HIST *The History and Social Relations of Genetics*, The Open University Press. This text is to be studied in parallel with the Units of the Course. We refer to it by its code, *HIST*.

mentioned several times previously during this Course. The phenylketonuric child shows progressive mental retardation from birth and in a normal environment does not live long. The condition was first diagnosed in the 1930s in Norway, when a dentist with two feeble-minded children noted that their urine gave off a peculiar smell. The substance was identified as phenylpyruvic acid, and it was suggested that the disease was due to a disturbance in the metabolism of the amino acid phenylalanine. Research showed that in affected individuals, the enzyme phenylalanine hydroxylase, which catalyses the conversion of phenylalanine to tyrosine, is missing; excess phenylalanine is instead converted to phenylpyruvic acid and excreted. Tyrosine is important in brain development because in normal individuals it is in its turn converted into substances that perform a transmitter role in the brain, conveying signals between nerve cells. This could be one possible link between the biochemical defect and the mental retardation that develops in affected children. But there are other ways whereby brain function may be damaged, such as the activity of toxic substances or the effects of intracellular chemical blocks resulting from the various metabolites formed following the primary defect.

How about the genetics of phenylketonuria? Figure 2 (on p. 623) shows a pedigree from a small population in an isolated group of small islands off the coast of Norway. Individuals marked with a cross were probably affected but died young. The pattern of transmission conforms to that expected of an autosomal recessive gene (note the marriage of cousins in the third generation). Of the 18 children in generation IV, 4 were definitely affected and 2 probably affected.

> **ITQ 15** If both unaffected parents of each sibship were carriers, then one-quarter of the children would be expected to be affected. The slight excess over expectation is not significant, but why do such pedigree charts frequently show an excess over expectation?

However, can we say that the PKU gene defect *causes* mental retardation? So far, we have been careful to speak of the effects of the PKU gene defect in children reared *in a normal environment*. If PKU is diagnosed at birth and the infant is placed on a diet lacking in phenylalanine, which is a normal dietary constituent, then the child develops more or less normally and *no longer* develops the mental retardation. This is a simple demonstration of the fact that any phenotype is the product of interaction between the genotype and the environment. We can only speak of the PKU gene as *causing* mental retardation *if we specify the environment in which that gene expresses itself*. In a changed environment, the behavioural phenotype is also different. Incidentally, the possibility of curing PKU by dietary change (an example of what has been called *euphenics*; see Section 15.4.4) has led to the development of mass-screening programmes for newborn children: a high level of blood phenylalanine may (though not necessarily) indicate the presence of the disorder. Nor does the enzyme deficiency automatically lead to mental retardation in untreated individuals; a significant proportion of older apparent phenylketonurics with around average intelligence has been revealed by mass screening—yet another example of the complex relationship of genotype with behavioural phenotype. It should be noted, of course, that the affected individual, even when cured, will still transmit the PKU gene to a subsequent generation.

euphenics

PKU is not the only disorder of its type; there are a cluster of related enzyme deficiencies in amino acid metabolism that are associated with mental retardation. There is also a group of genetically determined disorders of lipid metabolism, such as Tay–Sachs disease, metachromic leucodystrophy and the gangliosidoses, in all of which there is mental retardation, which varies in severity and in the age of onset. Abnormalities of carbohydrate metabolism such as galactosaemia may also have similar consequences.

However, for most of these disorders it is not easy to propose euphenic treatments. In addition, there are syndromes that appear, not during early development but in adult life. An example is *Huntington's chorea*, for which the mean age of onset is around 40–45 years and which is characterized by loss of motor control, and progressive impairment of mental processes. The pattern of transmission may be summarized as one in which (a) most sufferers from the disease had a parent who also showed it and (b) approximately half the children of an affected parent eventually develop the disease.

Huntington's chorea

> **ITQ 16** What does this pattern tell you about the mode of genetic transmission of Huntington's chorea?

14.3.5 Multiple-gene effects; schizophrenia

In the previous Sections we have introduced some of the particular problems associated with the study of behavioural genetics and have briefly discussed some examples of known human genetic disorders involving genes with major effects. The examples we have discussed, even those in which the biochemical and cellular aspects of the disorders are not well understood, are genetically relatively straightforward and non-controversial, through controversy may intrude when suggestions for social policy begin to flow from the genetic and medical observations, as we shall see in Sections 15.4.4 and 15.4.5.

More complex is the situation in which genetic tools are brought to bear on presumed behaviour where the diagnosis is dependent on criteria that are themselves hard to define and where there are no obvious cellular or biochemical abnormalities. This situation is exemplified by the attempts to assess a positive genetic contribution to the understanding of human behavioural disorders such as schizophrenia or depression.

'Mental disorders such as schizophrenia do not fall into a straightforward medical category, like, say cholera or a broken bone or even PKU; they cannot be diagnosed on the basis of physical signs or symptoms or biochemical tests but only on the basis of an individual's behaviour in interaction with family or doctor. To start with, there is the problem of the definition of a disease that is estimated to affect 500 000 people in Britain. Schizophrenia literally means 'split mind'. There are those who claim it represents merely an 'incongruity' or failure of communication between doctor and patient, but this is inadequate to describe the depths of feelings of those who are schizophrenic; they feel in some way fundamentally cut off from the rest of humanity, unable to express emotions or interact normally with others, or to express themselves verbally in a way that is rational to others; they appear blank, apathetic, dull; they may complain their thoughts are not their own or that they are being controlled by some outside force; dramatically ill schizophrenics appear not to be able to—nor wish to—do anything for themselves; they take no interest in food, sexual activity or exercise; they experience auditory hallucinations and their speech seems rambling, incoherent and disconnected to the casual listener. There are those who doubt that schizophrenia is a single entity at all, and the diagnosis of schizophrenia in a patient with a given set of symptoms can vary between doctor and doctor and culture and culture (comparisons of figures in different countries have shown that the 'widest' classification of schizophrenia occurs in the USA and USSR; none the less, even in Britain, where it is defined in a somewhat narrower sense, some 1 per cent of the population is said to be schizophrenic).

schizophrenia

A further problem in the study of schizophrenia has been the difficulty in achieving a common understanding among different disciplines of how the disease should be approached; extreme claims have been made that the schizophrenia diagnosis is exclusively the result of the interaction of the individual with the family and that the underlying biological background is irrelevant, or that the schizophrenia is inherited by way of a single recessive gene—or even a dominant gene with partial penetrance—or that *the* cause of schizophrenia is an abnormal metabolite excreted by schizophrenics in their sweat, or a dietary excess of wheat gluten.

It is probably safe to say that nothing useful, still less conclusive, is known about biochemical abnormalities in connection with schizophrenia. Any evidence concerning the possible inheritance of the trait must be assessed with no knowledge of the intervening variables between genotype and phenotype. Evidence suggesting a largely environmental origin for schizophrenia, accumulated especially following work in the 1930s in Chicago, indicated that the incidence of schizophrenia was highest in individuals in manual, working-class, occupations and living in derelict inner-city regions, and lowest in people in middle-class occupations living in the suburbs. This could be interpreted to suggest that schizophrenia results from the exposure to certain stressful types of living and working conditions.

ITQ 17 Can you suggest alternative explanations to account for the observed incidence of schizophrenia?

In fact, all of these possibilities—the drift of individuals, lack of consistence in diagnosis, *and* environmental stress—seem to occur, emphasizing the difficulty of 'mono-causal' explanations, at least for the 'global' disorder. Hence, although in the past there were those who believed that the 'cause' of schizophrenia might be located in the deficiency of a single enzyme or metabolic system, and thus be associated with a single-gene effect, the more sophisticated biological or genetic psychiatry of today would tend to look for more complex relationships.

The approach to studying disorders such as schizophrenia must be at many levels; there must be an epidemiological analysis of the distribution and incidence of the disease across time and social groupings; a psychiatric analysis, which tries to draw general conclusions from the detailed case histories of individual sufferers, and biochemical comparisons of metabolites suspected of having important roles in brain function between normal people and sufferers from the disorders. This last approach is made harder by the fact that the only tissues available for analysis on a long-term basis from living subjects are urine, blood or cerebrospinal fluid, which are all relatively remote from the brain, which must be the site of any biochemical lesion, if it exists. The contribution of the geneticist in this area is along the lines of pedigree analysis described in Section 14.1.1 and 14.1.2, including MZ and DZ twin studies. The techniques of somatic-cell genetics are largely irrelevant here.

The best tool of the geneticist in this context may be kinship correlation studies, and these indicate that, in the environment of the United States and of Britain, the chances of schizophrenia in children with one schizophrenic parent are about 12 per cent. Among the sibs of a schizophrenic the figure is 14 per cent, and with MZ twins the figure is claimed to be 90 per cent if they have shared the same post-natal environment, and 78 per cent if they have been separated for 5 years or more. Although the twin studies—particularly those involving twins reared apart—must be viewed with caution, it does seem likely that in any given environment the tendency towards schizophrenia is a consequence of the activity of many genes, possibly with some sort of threshold, so that relatively fortuitous environmental events may precipitate the person with a 'schizophrenogenic personality' from the latent to the overt disease. Such a general statement of the role of both genotype and environment in the onset of schizophrenia leaves open the question of whether the cure or alleviation of the symptoms of schizophrenia in an individual patient is best approached by the use of drugs that may act by altering a possible expression of gene action or may be used simply to deal with the symptoms, or by a modification of the individual's social and family environment, or by a concerted attack in both directions. The contribution of genetics to answering this social and medical question is bound to be limited. Even if the heritability of schizophrenia were shown to be high, this would not of itself mean that a biochemical treatment would be more appropriate than, for instance, psychotherapy. As we shall see more clearly in the following Sections, the potential of genetic knowledge as an aid in therapeutic intervention is very circumscribed.

Our chief example of the limits of the genetic approach to human behaviour comes in the case study of 'the genetics of intelligence and IQ'. However, because the discussion of the genetics of IQ has become inextricably bound up, in at least the public form of the debate, with the question of differences in performance, not between individual humans but between human racial and class groups, it is appropriate to postpone this study until after we have dealt with the genetic approach to human populations, the topic with which we begin Unit 15.

15.1 Genetic variation in human populations

It is commonplace that there is considerable phenotypic variation between people whether they are near neighbours or widely separated geographically. As with all other organisms, we can be sure that part of this variation is genetic in origin, and part is due to phenotypic plasticity. But how much of each? Is it possible to define different human populations or sub-groups in genetic terms, that is, in terms of the genetic differences between them? And if so, what relation do the genetically defined differences between populations bear to the political or social division of the world's human population into races, nationalities or ethnic groups? Because of the political and emotional charge that race, nationality and ethnic divisions carry, and because these terms and divisions were in use long before the advent of modern genetics, but are sometimes assumed to be genetic in origin, these questions are of more than 'purely' scientific concern.

human genetic variation

The questions that will be dealt with in this Section are:

1   What is the extent of variation in human populations (Section 15.1.1)?

2   Is it possible to define different human groups in terms of their genetic constitution, and what is the genetic status of the concept of race (Section 15.1.2)?

3   How may genetic differences among different human geographical groups have arisen (Section 15.2)?

In this part of the text we assume that you have made yourself familiar with the basic concepts of population genetics in Units 9–13; in part, some of the examples we use to illustrate principles here have already been demonstrated to you in other species.

15.1.1   The extent of human genetic variation

The number of human genes can be estimated only indirectly. One type of method is based on the comparison of the frequency of mutations of known genes with the total number of all mutations that are seen to occur. If the over-all rate of mutation is constant then the likelihood that any given gene will show a mutation will depend on the total number of genes present. Calculations using this method show that the number of human genes is of the order of tens of thousands.

The catalogue of known mutant human genes is much smaller than tens of thousands, but is steadily increasing. McKusick has described the growth in knowledge of severe phenotypic effects caused by mutation in man: in 1966, 1 450 different mutants had been identified and classified, each one of which caused what is known as a 'genetic abnormality'. In 1968 this figure was 1 550 and in 1975 it had risen to over 2 300. Of course, the increase in these totals reflects advances in recognition and not a steadily increasing rate of mutation! The total frequency of appearance of severe genetic defects (for example, phenylketonuria, hereditary dwarfism or defects in other single genes or chromosomes) is now estimated at about 1 per cent among live births (though 2 per cent is an often quoted figure). Many more infants have a variable genetic predisposition to disease. In addition, perhaps 10 per cent or more of child deaths in Britain are estimated to result from genetic diseases. At least 30 per cent of spontaneous abortions—which represent more than 15 per cent of human conceptions—have a major chromosome anomaly (for example, an autosomal trisomy, sex-chromosome monosomy (XO) or triploid; see Unit 5, Section 5.2.2). Also many chromosomally normal embryos and fetuses with serious developmental anomalies (presumably related to an underlying genetic predisposition) are spontaneously aborted so that, proportionally few survive to term.

In this part of the text we are concerned with the total extent of genetic variation, but we shall return to these figures later, as they represent an appreciable problem for medicine and society (Section 15.4.4).

The phenotypes of 'genetic diseases' are extreme and obvious, but, in addition, there are many other more subtle phenotypic differences due to gene differences, which affect the predisposition or relative resistance to disease, as well as many phenotypic differences that may not have adversely affected the health or normality of the individual.

ITQ 18   What would be a simple method of detecting small changes in protein structure that may reflect genetic differences?

As you will recall from Units 9 and 10, the method referred to in ITQ 18 has been applied to the detection of polymorphism in both humans and other species; the study of Norris and his colleagues on isoenzymes in the European population, in which 71 loci in all were studied, found that 20 showed polymorphism. This represented 28 per cent of the total loci, giving an average heterozygosity (Radio programme 9) of 0.067, which means that any individual is likely to be heterozygous at about 7 per cent of the gene loci that code for enzymes. However, only one-third of all possible amino acid changes in proteins give rise to a change in charge that can be detected by electrophoresis. So the actual heterozygosity may be three times that observed, or 20 per cent. A comparable figure was obtained on analysis of the genes that code for the antigens on red blood cells. By 1967, 33 different genes determining antigens had been defined by antigenic tests, giving an average heterozygosity of 0.162, and the total proportion of polymorphic loci was 36.4 per cent. The generality of these conclusions requires the sample of genes being examined to be typical and not unusual in any way, and depends on the discrimination of the biochemical methods involved. Both clinical and biochemical studies thus confirm the considerable genetic variation in human populations. Using the techniques of electrophoresis and antigen–antibody tests, genetic polymorphisms (Units 9 and 10, Section 9.8.1) have been discovered for a large number of proteins identifiable in individual blood samples. The majority of these allelic differences are recognized, as we saw in the rhesus system, by their agglutination reaction in mixtures with standard antisera prepared against pure samples of particular proteins.

15.1.2 Genetics and the concept of race

This Section is devoted to an explanation of genetic diversity within and between different human populations. Our examples involve such groups as the North American Blacks and Indians, Eskimos and the Jews. It is important to distinguish two very separate uses of the word race. In all its uses, race refers to a group of people identified by some common features or attributes. The problem so far as this Course is concerned is to be clear about whether and when the word is being used in a genetically meaningful way, because many confusions and misconceptions arise when it is not clear precisely what definition of the word race is being employed.

The biological use of the term race—as applied to all species, not just humans— normally refers to a breeding group or population (possibly in a distinct geographical area), within which there is free exchange of genes, and which may be identified as a group by some common and distinguishable heritable attribute.

A breeding group can be defined genetically by the composition of its gene pool (Unit 12, Section 12.6). One useful way of identifying genetic differences between individuals is by way of blood-group differences; the analysis of blood protein is a clear and largely unambiguous method. Yet even its use does not allow unequivocal allocation of any individual to a particular race. When one describes a gene pool as, say, consisting of 10 per cent I^A, 10 per cent I^B and 80 per cent I^O alleles (these, you will recall, are the ABO blood-group alleles), one is specifying population parameters. Within that population there will be individuals homozygous and heterozygous for the various alleles. Right away you will see that a definition of 'race' in terms of particular allele frequencies means that it may be impossible to say with certainty whether any chosen individual belongs to that race or not (no individual can, for example, have 10 per cent I^A, 20 per cent I^B and 90 per cent I^O!). With more complex systems than this, such as skin pigmentation, in which genetic control and phenotypic plasticity are much more important, the difficulty of 'typing' any individual becomes too great to be genetically useful. In addition, the concept of a 'race' defined by its gene pool and its allele frequencies, is a dynamic one that may be changing with each generation in response to forces that tend to change allele frequencies.

> ITQ 19 Which of the events listed below (a) necessarily will lead to changes in allele frequency, (b) may under particular conditions lead to changes in allele frequency and (c) never lead to changes in allele frequency?
>
> (i)   selection against one homozygote
> (ii)   migration in and out of the population
> (iii)   mutation
> (iv)   genetic drift in small populations
> (v)   assortative mating (departure from random mating)
> (vi)   recombination
> (vii)   inbreeding

In contrast to the genetic definition, the social definition of race depends on social ascription, based on real or presumed cultural or physical differences; these differ from society to society and time to time. For example, in the 1930s there was a vogue (now discredited) among eugenicists for a division of Europeans into three 'races', so-called 'Nordic', 'Alpine' and 'Mediterranean'. Today in South Africa, Japanese are classified as 'honorary whites', but Chinese are 'coloureds'. Both the social characteristics ascribed to Jews and the legal definition of what constitutes a Jew has differed between, say, Germany in the 1930s and Israel today. For such social definitions an 'easily recognized' characteristic like skin colour is necessary, and the genetics becomes largely irrelevant. As soon as the definition includes cultural or political components, tacitly or openly, the race concept no longer bears much resemblance to the biological definition. To claim that there was complete agreement on both definitions one would have to show that the cultural or political groups were co-extensive with distinct gene pools, and had an identifiable biological and genetic reality. The German anthropologist, Kossinna, avoided such problems by stating that 'Nordic souls may often be combined with un-Nordic bodies, and a decidedly un-Nordic soul may lurk in a perfectly good Nordic body'!

Examine the data of Table 11 (Lerner, 1968) concerning the allele frequencies for the ABO blood group in two geographically, culturally and politically distinct populations, the Basques of Spain and the Asian Mongoloid population.

Table 11 Allele frequencies in the ABO blood group in Basque and Asian Mongoloid populations

Population	Frequency/%	
	I^A	I^B
Basque	23	0–3
Asian Mongoloid	15–25	15–30

The two populations clearly differ very markedly, and on the basis of this evidence (although one blood group alone is thin evidence, and one would have to show that a similar partition occurred for alleles at many loci), one would accept that they are genetically distinct races.

But now look at the data of Table 12 (Roberts, 1942), taken from North Wales in the 1940s. Note that the differences between the allele frequencies for the two groups were found to be statistically significant.

Table 12 Percentages of various phenotypes for the ABO blood-group system among blood donors with Welsh and non-Welsh family names

Men and single women donors	Numbers of individuals	Frequency/%			
		I^O	I^A	I^B	I^{AB}
Welsh family names	909	52.7	35.0	9.7	2.6
non-Welsh family names	1 091	46.6	42.0	8.3	3.2

Thus, there are clearly genetic differences between these two groups identified on the basis of their ancestry, despite the fact that they live closely together and that no one would claim that the two formed separate races.

15.2 The origins of genetic differences between human populations

This Section contains a number of examples of genetic differences between human populations, which have been demonstrated in terms of differences in allele frequency. The particular populations mentioned will give you an opportunity to examine your ideas on the concept of race, but the main intention is to show what can be concluded about the origins of these differences in allele frequency. Our conclusions, therefore, will be hypotheses about the possible genetic history of these groups, and you may wish to compare these hypotheses with information on human social history, derived from other sources, including archaeology and anthropology.

The Navajo Indians of the south-west and the Eskimos of the north of the USA are two very distinct social groups. Table 13 (Lerner, 1968) gives the allele frequencies for the ABO blood-group polymorphism and for the MN blood-group polymorphism for each population.

Table 13   Allele frequencies in Navajos and Eskimos

Blood-group polymorphism	Allele	Allele frequencies	
		Navajo	Eskimo
MN	L^M	0.917	0.913
	L^N	0.083	0.087
ABO	I^A	0.013	0.333*
	I^B	0.000	0.027*
	I^O	0.987	0.6420*

* Significantly different in the sample size studied.

ITQ 20   On the basis of the data in Table 13, do you think the two groups are closely related genetically?

This illustrates the first point to be made, namely, that comparisons made for one gene alone are not going to give very reliable answers. In this instance we would need to pursue the comparison using a larger number of genes to see which, if any, of the genes compared was most representative of the difference between the populations. If the two populations turned out to be very similar for most genes examined, then one would conclude that they were quite closely related populations. That would still leave us with the problem of explaining why allele frequencies for the ABO system were so different.

Now let us take the analysis further and consider a comparison of several different human populations in order to see if they can be grouped by their genetic 'relatedness' so as to represent a *phylogeny* (or evolutionary relationship; see Units 9 and 10, Section 10.6.2), a grouping that reflects the evolutionary origins and divergences that may have occurred. Here we are deliberately setting out to try to identify the direction of systematic and directional changes in allele frequency, and, by comparing such differences, to decide on the 'family tree' of the particular populations. It follows that the comparisons should be based on genes that are polymorphic in the populations. (You will meet the use of rare genetic differences shortly, in another context.) So the first thing to be established is that the comparison will be based on as many poly-morphic genes as can be conveniently classified in the various populations. It has further to be assumed that the alleles are not under strong selection, and certainly not under different selection pressures in the different populations. Finally, a decision has to be made about how to measure 'relatedness' so as to be able to construct a map or 'tree' of relationships.

phylogeny
evolutionary divergence

The logic is of the following form. Suppose we have three populations A, B and C in which the *M* allele frequency of the MN blood-group polymorphism is 10 per cent in A, 50 per cent in B and 30 per cent in C.

The simplest relationship between populations A, B and C is that A is more closely related to C than to B, so that we could represent the relationship as A———C———B, in which the lines join the most closely related populations. So you may imagine how the frequency of an allele for one gene would allow you to order or rank the populations in a linear fashion according to similarity. For two and for three genes, graphical plots in two or three dimensions will suffice, but somehow a measure had to be developed that included the differences at the many polymorphic loci measured to give an estimate of total relatedness. Edwards and Cavalli-Sforza designed and carried out such an analysis for 15 populations in which they looked at the allele frequencies of 5 genes. Several maps of the relatedness of the 15 populations could be constructed, all of which were equally good fits, but one example is shown in Figure 14 in which you should note particularly the branch points. The branching network has been superimposed on a map of the world in which the closed circles indicate the locations of the 15 populations.

QUESTION   What implication or possible explanation is there behind the fact that the branches of the network have largely been drawn over land, even when this is not the shortest distance between two populations?

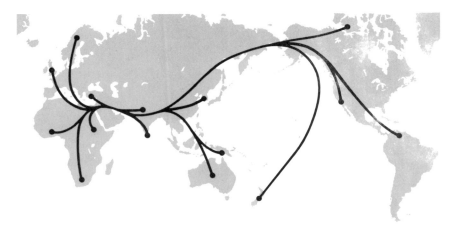

Figure 14 World map of allele frequencies of 5 genes (after Edwards and Cavalli-S   1964).

ANSWER The strong implication is that the relatedness means that there has been or is genetic contact, or gene flow, between these populations and that this contact has largely been by people moving over land, rather than by sea.

The connecting branches do happen to fit closely with believed migration routes in human history: through Egypt; across the Middle East to India; the Bering Straits; from the west coast of the Americas to the Pacific Islands; and so on. If the populations have been derived from these migrations, with a concomitant gene flow, genetic relatedness would be expected. A related study on a micro-scale was carried out for the populations of a number of mountain villages in northern Italy, and the analysis of relatedness between the villages was drawn as the 'best-fit' map of inter-relations between the villages. It turned out to mirror the road map for this region. Not a blindingly new insight, you may be thinking. But it seems reasonable that the genetic relatedness could reflect the 'migration' of alleles from parish to parish as marriages were contracted between people of neighbouring villages. One might be able to verify such a model by actually looking at the marriage registers of the villages to see if the mobility is as predicted by the genetic model.

Such analysis could possibly be applied in a wide range of circumstances. It could be used to help decide whether a particular theory of the origin of an island population holds up, or whether various cultural and linguistic similarities between geographically separated groups can ever be 'explained' by their common ancestry.

ITQ 21   Let us suppose that from an analysis similar to those above, two geographically isolated populations are discovered to have very similar gene pools and also have very similar and unique features in their spoken language. Which of the following conclusions would you reach?

(i) This is evidence for a heritable component in the form of the language.

(ii) The adoption of similar languages favoured particular alleles and the two populations 'converged' in genetic terms under selection.

(iii) The two populations had a common biological and a common social ancestry.

This hypothetical example serves as a reminder of the assumptions and constraints in this method. They include:

1   The phenotypes used to estimate allele frequency must be largely unaffected by possible environmental fluctuations.

2   The total measure of genetic differences is a good measure of possible evolutionary divergence.

15.2.1   The Amish and the North American Indians

The Old Order Amish is a religious sect represented by a number of communities in Pennsylvania, Ohio and Indiana in the USA, and in Canada. These groups are distinct and marriage is largely within the group. The frequency of four autosomal recessive genotypes, all of which lead to deleterious phenotypes of varying degree, are very high in these groups. For example, it is estimated that 13 per cent of the Amish in Lancaster County, Pennsylvania, are heterozygous for the polydactyly allele. In the rest of the United States population all four phenotypes are rare. This sect originated in Europe and groups of Amish migrated to the USA.

> **ITQ 22** Choose the most likely explanation from those below for the high frequency of these alleles. You may find it helpful to recall the major factors affecting change in allele frequencies in populations, which are selection, differential migration of particular genotypes, genetic drift and mutation.
>
> (i) In what were probably small initial groups of settlers (founder groups), strong selection was operating in favour of these recessive homozygotes.
>
> (ii) Small initial groups of settlers may have by chance brought atypical samples of the alleles from the ancestral population in Europe.

This general phenomenon is termed genetic drift (Units 9 and 10, Section 10.5.1), or, perhaps more precisely in this example, the founder principle (Units 9 and 10, Section 10.5.4).

The majority of the North American Indian tribes have a very high frequency of allele I^O in the ABO system, but there are exceptions. The Blackfoot and the Blood tribes have about 80 per cent I^A alleles. The composition of the gene pool of these two tribes may again be a result of random genetic drift, either because a preponderance of O phenotypes colonized the continent and the rare A types were among the few founders of these two tribes, or because even after the initial establishment of the tribe the small population size meant that random fluctuations could alter the allele frequencies from generation to generation.

15.2.2 Black populations in the United States

The Black population of the USA derives from Negro slaves brought from West Africa, the majority of the White population are derived from European emigrants. (Some people with some Black ancestors may be regarded as 'White'.) Even though there are still black and white-skinned individuals generations after the arrival of these groups, have these two groups preserved the genetic differences, other than those affecting skin pigmentation, that the two ancestral populations undoubtedly had? Skin colour is still an important issue in political and economic terms in the USA, but is it really an indicator of the genetic separation of the two 'races'?

An approach to answering this question was given in Units 9 and 10, Section 10.4.2.

> **ITQ 23** From your work on those Units, name the factors other than the degree of 'mixing' that could alter allele frequencies in the Black population.

You will recall from Units 9 and 10, Section 10.4.2, that when actual studies were made on allele frequencies for the blood groups G_M and Duffy Fy^a, in the 'ancestral' population, European Whites and West African Negroes, and in Blacks from Oakland, California, between 22 per cent and 28 per cent of the genes in the Black population were found to derive from the American White population, which could be accounted for simply by admixture. On the other hand, if the haptoglobin Hp^1 or the sickle-cell haemoglobin Hb^S alleles are considered, the figures come out at 30 per cent, which is higher than could be accounted for by admixture alone and is likely to indicate that selection has also occurred.

15.2.3 The Jewish biological race or the Jewish culture?

Jewish communities have maintained their identities throughout the world in many different populations. When Jews from various countries moved to Israel to take up Israeli nationality, the opportunity was taken to investigate the composition of the gene pools of the Jewish communities around the world, of which these individuals were representative. Frequencies of the *G6PD* allele, that is, the recessive allele that determines the inactive form of the enzyme glucose 6-phosphate dehydrogenase, varied with the origin of the Jewish individual, as shown in Table 14 (Lerner, 1968). In addition, the ABO blood-group allele frequencies also varied according to the origin of the Jewish population, yet showed close agreement with the non-Jewish population of the area from which they came.

Table 14 The frequency of the *G6PD* allele in male Jews from different areas

Origin of Jewish male	Frequency of *G6PD* allele
Kurdish	0.60
Persian and Iraqui	0.25
Turkish	0.05
Yemenite	0.05
North African	0.02
European	0.002

ITQ 24 From the above information and the discussions earlier in these Units, can you eliminate any of the following as incorrect conclusions?

(i) The Jews are a distinct race in the genetic definition of the word, even though for many generations they have been physically separated in different parts of the world.

(ii) The fact that Jewish and non-Jewish populations in the same area have similar allele frequencies is confirmation that natural selection has modified two adjacent and separate populations in the same way.

(iii) The similarity in allele frequencies between Jewish and non-Jewish populations of the same location indicates that gene flow and admixture has occurred.

It is a fairly general part of 'lay' belief that socially ascribed races or ethnic groups do not mix and rarely interbreed. The genetic evidence given in this Section suggests, on the contrary, that gene flow between populations that have obvious physical (colour) or socio-religious distinctions does occur, on a quite considerable scale.

15.2.4 The distribution of the ABO blood-group alleles in Europe

Figure 15 (p. 656) indicates the cline (Units 9 and 10, Section 9) for the blood-group allele I^B in Europe, showing a high frequency in the south-east gradually falling off to a frequency of below 5 per cent in parts of Spain and Portugal, in the west.

Consider the three following possible explanations for this cline, in particular the assumptions you need to make and consequences that flow from adopting each explanation: (a) random genetic drift, (b) selection and (c) migration or mixing.

Let us consider each alternative in turn.

(a) *Random genetic drift* This is not probable, as it is rather unlikely that a dispersive process such as drift would produce such a regular gradation in frequency of the I^B allele across the continent. Instead, one might expect such a randomizing process to generate local peaks or 'islands' in which allele I^B was high, separated by areas in which it was lower. In addition, one would expect some relationship between the extent of drift, or founder effect, and population size at the time, which ought to relate in turn to the opportunities that particular areas of Europe offered for successful settlement.

(b) *Selection* This would imply that in the areas that contained a low frequency of I^B (the more westerly areas) there was stronger selection against this allele, or, conversely, that allele I^B was favoured over I^A and I^O in the eastern continental area. One would, then, have to suggest some environmental factor that correlated with the map of the I^B allele frequency. For example, one might consider climatic differences, the distribution of diseases like smallpox or plague, nutritional changes or different social customs that somehow might predispose particular genotypes to selection. However, any such correlation might be fortuitous, and would still not imply a causal relationship unless backed up by a good deal of further study.

(c) *Migration or mixing* On this explanation, the contours of frequency of the I^B allele would represent the effects of a pattern of migration or mixing, from east to west or from west to east. This distribution could be the result of the invasion by Tartars and Mongols from the east from 500 to 1500 AD bringing a higher frequency

655

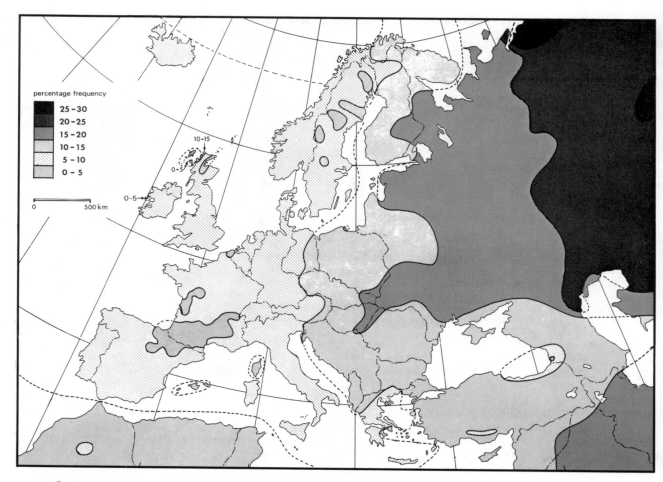

percentage frequency

■	25–30
■	20–25
▨	15–20
░	10–15
╱	5–10
▒	0–5

0   500 km

10-15

0–5

0–5

of the I^B allele to a Europe in which the pre-existing frequency was low. Local differences in, for example, the Iberian peninsula and in central Sweden might on this explanation represent populations that were largely isolated from the migrants from the east.

Figure 15 Cline for the blood-group allele I^B in Europe (drawn by John Hunt).

Which of these explanations is the more correct is not answerable with just the data you have been given, and you will realize that it might be very difficult to establish causal connections between, for example, resistance to infectious diseases and a person's ABO blood-group genotype. In addition, one would require further information on the size, distribution and migration behaviour of the population of Europe, before the hypothesis of genetic drift could be investigated. It is interesting that there is also a strong cline in the frequencies of the same genes within Britain itself from Scotland down to East Anglia and from East Anglia westward into Ireland. The relationship of these clines to historical facts of invasion and migration is striking.

15.2.5   Frequencies of rhesus alleles in the population; the effect of gene flow

The Units on population genetics (Units 9–12) have introduced you to the idea that evolution and genetic divergence *may* follow the geographical or other isolation of two formerly genetically similar populations. Subsequent breeding between members of the separated populations may give rise to problems as the consequence of differences in chromosome arrangements (Unit 5, Section 5.4), or perhaps more subtly as the result of the formation of heterozygotes derived from matings between individuals carrying genomes that have become separately adapted to different environments. In humans, what are the consequences of gene flow between largely isolated groups? The example of the Amish demonstrates one of the undesirable outcomes of inbreeding in a closed population: the increase in the frequency of homozygotes, among which will be some more or less deleterious phenotypes. In such an instance, outbreeding would reduce the frequency of homozygotes and the consequent risk of the appearance of deleterious phenotypes. On the other hand, up until the last ten or so years, gene flow between western European and Asian populations had brought problems associated with rhesus haemolytic disease of the newborn.

gene flow

Although the reasons for the variation are unknown, it is apparent from Table 15 (Mourant *et al.*, 1976) that Asian, American Indian and African Black populations

Table 15 The distribution of *Rh* alleles in different human populations

	%*Rh*$^+$	%*Rh*	Sample size
Basques	69.55	30.45	1 573
French (Montpellier)	81.19	18.81	2 403
English (Nottingham)	82.32	17.68	5 000
Scottish (Aberdeen)	82.81	17.19	3 601
Jugoslavia (Belgrade)	83.81	16.19	20 000
Blacks (Baltimore USA)	92.83	7.17	907
Bantu (Kisangani)	95.20	4.80	4 000
Katangese	96.75	3.25	400
Chinese (Peking)	99.40	0.60	2 324
Japanese (Tokyo)	99.56	0.44	4 541
Eskimos (W. Alaska)	99.96	0.04	2 522
Chippewa Indians (Minnesota)	100.00	0.00	161

have 95–100 per cent of the *Rh*$^+$ allele, whereas in Europe the frequency of *Rh*$^+$ is around 80 per cent. Thus, this form of haemolytic disease is practically unknown in Asian populations. However, if Europeans introduce *Rh* alleles into Asian populations and inbreeding follows, then this disease will start to appear. Conversely, the introduction of genes from Asian populations (largely *Rh*$^+$ alleles) will increase the frequency of this allele and decrease the frequency of homozygotes of genotype *Rh/Rh*. Thus gene flow would seem to create problems for the Asian populations, but it improves the situation for the European population. (But remember that we are talking about biological and not social ascriptions of race; it cannot be emphasized too strongly that there are no conceivable 'social policy' conclusions that can properly be drawn from such observations.)

15.2.6 Genetics and the concept of race

We can conclude from the last two Sections that although genetic studies can perhaps cast some light on the evolution and separation of human populations, and the study of human genetic polymorphisms will doubtless continue to be a fascinating research topic for many years, none the less neither geographical, cultural or political separation or contiguity are clear predictors of genetic differences or similarities, and vice versa. The study of human genetics is not aided by attempts to decide which groups or individuals are or are not members of which races, nor are social and political questions about the integration or segregation of particular groups ('ethnic' or 'racial' in the social sense) to be answered by any appeal to the findings of genetics.

Before the advent of genetics, the literature was full of discussion and dispute about which geographical groups were races, the classifications being based in the main on conspicuous physical characteristics such as skin colour, build and facial characteristics (with presumed heritable components). If you decided on few races you were a 'lumper' and if you had more classes than everyone else, a 'splitter'. It is difficult to imagine what meaning the idea of a 'pure race' has in human terms, as for most of the alleles that have been studied there is very much greater variation *within* socially defined 'races' than there is *between* such races. If 'pure' is taken to mean homozygous, then it is obvious that the biological significance of the term is very limited as only in certain domesticated species such as cereals and some laboratory animals does one find completely or at least highly homozygous stocks. Thus, the concept of a 'pure race' is strictly political, and has no biological significance. The conclusion has to be that a strict genetic use of the term race is rather far removed from the colloquial use of the word. Geneticists nowadays prefer to talk of humans as *polytypic species*, meaning that there is no such person as a 'typical' human but that there are many different alleles and genotypes, giving rise to phenotypic differences, widely distributed, but not exclusively confined to any sub-group or geographical area. These differences themselves may also be confounded with environmental effects such as climate, diet and disease, which in their turn may help maintain some of the differences of allele frequencies (see Section 15.4).

polytypic species

'Race' as a concept arose before genetics and there was a time when individuals of different 'anthropological' races were characterized by claims that they possessed particular obvious phenotypic differences either physically, or in behaviour, or both; such claims were characterizing a 'stereotype'. Such a norm or stereotype is completely at odds with the manifest variation that occurs in most characteristics in biological populations. Even if it did exist, it would be no more than a statistical concept, which could never apply to every individual or help to identify any one person as 'belonging to' one or other race. One should perhaps instead question the intent behind such classifications, being clear as to whether they are biological or social (noting at the same time that the social classification of race is often used *as if* it had the implied authority of biological meaning).

15.3 Genetics and intelligence

We have now assembled the necessary background—on the techniques of genetic investigation in human populations, on the geneticist's, as opposed to the 'popular', use of the word race and on the particular problems that beset the study of behavioural genetics—to approach the final case study of this Course, which is concerned with the possibility of investigating the genetic and environmental contribution to the differences in intelligence among individuals or between social and 'racial' groups. Discussion of this issue, sometimes known as the 'race–IQ debate', has created dissent, often acrimonious, not just in the closed world of human geneticists and psychologists but also in much wider social and political circles. It is therefore appropriate to look *both* at the genetic and psychological methods, theories and interpretations that are available and *also* at the wider social framework within which they are set, for neither can be well understood without the other. Although most controversy has centred around the IQ case, it should be pointed out that the arguments have been extended by the protagonists of what has become known as the 'hereditarian' school (notably Eaves and Eysenck) to claim a substantial genetic component in a very large range of human social activities, including, for instance, political attitudes (radicalism versus conservatism), or even such characters as 'social ease' and 'conversational poise'!

15.3.1 The re-emergence of an old debate

The belief that (a) intelligence is inherited and that (b) there are genetically determined differences in intelligence among social classes, sexes and races has a history as old as that of genetics itself. However, for reasons to which we shall return below, for many years, from the 1940s to the late 1960s, the issue was believed to be resolved once and for all, along the lines of a UNESCO report (Montagu, 1972) that appeared in 1951 and that concluded:

> according to present knowledge, there is no proof that the groups of mankind differ in their innate mental characteristics whether in respect of intelligence or temperament. The scientific evidence indicates that the range of mental capacities in all ethnic groups is much the same.

Despite some attempts to reopen the question in the years that followed, it lay dormant until 1969, when an article was published entitled 'How much can we boost IQ and scholastic achievement?' by A. R. Jensen, a Californian educational psychologist. Jensen started from what he regarded as the failure of the compensatory educational programme in the US to achieve the educational 'catch-up', which had been widely assumed would occur among the ghetto children who participated in the programme. He reviewed the literature on educational performance and the genetics of IQ differences between individuals and groups, particularly in relationship to US Blacks, and concluded:

> There are intelligence genes, which are found in populations in different proportions, somewhat like the distribution of blood types. The number of intelligence genes seem lower, overall, in the black population than in the white.

> Jensen, 1969

The article created an immediate furore. In the USA Jensen's lead was followed by a number of others, notably the psychologist Herrnstein, who extended the analysis to claim that most of the determinants of class society were also due to genetic differences, and the physicist Shockley who derived the 'logical' policy conclusions by noting that, as working class Blacks and Whites tend to have larger families than the White middle class, it followed that the national intelligence was declining. Shockley's policy recommendation was for a programme of cash induce- ments for sterilization linked by a sliding scale to the sterilizee's IQ score. For Shockley,

> Nature has color-coded groups of individuals so that statistically reliable pre- dictions of their adaptability to intellectually rewarding and effective lives can easily be made and profitably be used by the pragmatic man in the street.
>
> Shockley, 1972

In Britain, the strongest support for Jensen's views has come from the psychologist, H. J. Eysenck, who in 1971 set out the case in the popular book *Race, Intelligence and Education*. By this time the issue had long ceased to be (if it ever had been) 'purely' academic; it was not only Shockley who drew policy conclusions, but also United States segregationists, and those in Britain who noted the high proportion of children from West Indian families in ESN (educationally subnormal) schools, drew sustenance from the case put forward by Jensen. Organizations were set up to campaign against what had become known as Jensenism, academic meetings and debates were picketed, newspaper editorials demanding freedom of scientific research were published, and many geneticists were called upon to take public positions in a way that had not happened since the 1940s (see *HIST*, Section H.4). The question of whether any particular scientist who has become involved in this discussion is or is not 'racist' is quite irrelevant to these Units; what we must examine are the theories and evidence used to support the claims being made. However, this cannot be done in isolation from the social and political ramifications of these views—the history of IQ testing and of eugenics.

A further point. It would be wrong to pretend to a 'dispassionate view' on this question—if indeed one could exist. Members of the Course Team who prepared this Unit are not above 'the battle of ideas' with which it deals, and we come down to a firm conclusion. If you wish to read other contributions to the discussion, you are encouraged to do so, and will find a short list in the Bibliography*. However, it is important to stress from the beginning that the debate is *not*, despite claims often made to the contrary, between a 'hereditarian' group on the one side and an 'environ- mentalist' group on the other, although these terms have been used in popular writings (and indeed taken up by sociologists of science writing about the issue). Although it is true that some of the so-called 'hereditarians' dispute most vehemently that 'the environment' has more than a very small contribution to make to the determining of intelligence, the majority of their opponents argue, as we shall see, that the methods of genetics and psychology have been incorrectly applied to give this seeming result, and that although genotype and environment must both contribute to any individual's 'intelligence', estimates of IQ scores and of their heritability cannot say anything meaningful about the reasons for differences in performance among individuals or groups.

'hereditarian' thesis

In order to review the claims that are made by Jensen and by others, it will be convenient to summarize them as a set of propositions. The general concern is to ask the question: what proportion of the differences in intelligence between individuals within groups on the one hand and between groups on the other is genetic, and what proportion is environmental?

The propositions run as follows:

1 There is a human character, intelligence, that can be measured by IQ tests.

2 In Britain and the USA, working-class people, Irish, Blacks and Mexican Americans, score lower on IQ tests than do middle class, British and White Americans.

3 Studies on the heritability of IQ within the White population suggest that 80 per cent of the variance between individuals can be parcelled out as genetic, 20 per cent as environmental, with a negligible G × E (genetic × environmental) interaction component.

4 For the purposes of applying these calculations to social groups, Blacks, Mexican Americans, etc., can be regarded as representing biologically defined as well as socially defined races.

5 The mean differences between these groups are larger than can be accounted for by the 'environmental' factor, and hence are genetically based.

Let us now review each of these propositions, and the evidence for them, in turn.

* Other recommended books are listed in the *Introduction and Guide to the Course*. (The Open University (1976) S299 IG *Introduction and Guide to the Course*, The Open University Press.)

15.3.2 IQ tests and intelligence

The first problem with which the student of the biology and psychology of human behaviour is faced in this area is obvious: is any sort of global assessment of human 'intelligence' possible? Common parlance speaks of people as 'intelligent' and rates one person as 'more clever' than another, and, indeed, the entire examination system as it has evolved in educational institutions in most advanced industrial countries is devoted to deriving quantitative methods for ranking people on scales of performance.

If one person is better than another at, say, carpentry or studying biology Course Units, will they also be better at mathematics or music or football? Which of the vast range of human activities shall we include in our scales for intelligence? Still further complexity is introduced by the fact that 'intelligent behaviour' is generally expressed not by an individual in isolation, but in interaction with other people, who are either cooperating or competing with or evaluating him or her. These are very formidable obstacles in the way of recognizing and measuring 'intelligence' as a phenotypic character.

These problems have beset human psychology from the outset. None the less, it has generally been assumed, especially among British and American psychologists, that a quantification of intelligence is possible and that individuals can in principle be ranked for intelligence on a linear scale. Some of the earliest attempts to examine the inheritance of intelligence were made by Darwin's cousin, Francis Galton, in the 1860s and 1870s. Galton (*HIST*)was the founder of biometrics and of the movement for genetic planning of human breeding—eugenics. His approach to the question of intelligence was, rather than to quantify it directly, to try to identify obviously talented individuals in the population, and to collect pedigree charts for them, so as to see whether talent 'ran in the family'. In *Hereditary Genius*, Galton (1869) studied the relations of a variety of eminent men (1 in 4 000 of the population of Victorian England, he estimated, fell into this category) and showed conclusively that judges, statesmen and divines, literary men and scientists tended to have among their relatives, often stretching back through several generations, other judges, statesmen, etc. Here, Galton concluded, was incontrovertible proof that genius was inherited, and of inherited genius, Britons above all, and other European 'races' to a lesser degree, were disproportionately endowed compared with other 'races'. (Note that Galton's use of the term race was in the social rather than the biological sense discussed earlier, but his arguments imply that the two definitions are coterminous and that it is important to identify group differences in intelligence.)

Galton's judgement of talent and genius was subjective and unquantifiable. Quantification was introduced by the work of Binet, in Paris, in the early years of the twentieth century. The French Minister of Public Instruction had commissioned him to identify students whose academic performance was so low as to necessitate placement in 'special schools', and in 1905 the first intelligence test was devised in the form of a series of questions that individual children were to be given, designed to give a high correlation with school performance. Their scores on the test could then be compared with the average scores for all children of the same age in the population to produce a measure for predicting school success. The term IQ or *intelligence quotient*, represents the ratio between the child's 'mental age' as measured by the tests, and his or her chronological age. For the average child, this ratio is unity and is represented as 100, and the test items are selected in such a way that the scores of the population of children being tested are distributed normally around the mean of 100 and the standard deviation of the population is defined as 15.

intelligence quotient (IQ)

This method of determining IQ is appropriate only for children; when applied to older individuals the relationship between 'mental' and 'chronological' age ceases to be meaningful; for adults, when the tests were later developed in the USA, special extensions of the test were included.

Binet regarded his tests as diagnostic and therapeutic tools; he believed that children's performance could be modified by 'mental orthopaedics', and to those who claimed the intelligence of an individual was a fixed quantity, he replied, 'we must protest and react against this brutal pessimism'.

Binet's tests were brought to Britain by Burt, and formed an integral part of the testing procedure for secondary school selection (the 11 + examination). In the USA, the tests were modified by Terman at Stanford and Goddard in New Jersey, and the now classic Stanford–Binet scale was first published in 1916. Unlike Binet, the new generation of testers was convinced that the intelligence they were measuring was a fixed quantity, a biological property of the individual, which could only be slightly modified by the environment. In the 1920s, Spearman proposed that, underlying all such tests was a 'general intelligence factor', or g; IQ tests measure g and the words intelligence and IQ came to be used virtually synonymously. For the educational psychologists of the 1920s and for some today, intelligence is essentially *defined* as 'what IQ tests measure'. All of the various components of potential academic performance are subsumed into the one category of IQ.

Let us look more closely, then, at the nature of the tests. There are several varieties now available, all testing the ability to manipulate some combination of figures, numbers and words. Tests that rely on non-verbal skills and that present items not easily related to general knowledge are termed 'culture-free' or 'culture-fair' and in theory they should be equally difficult for any person, no matter what their background. Thus many of the standard test items involve matching, mathematical or logical and linguistic skills, which, it is argued, should not discriminate on the basis of school or home background. None the less, one commonly used test, the Stanford–Binet, includes such questions as 'What is the thing for you to do when you have broken something that belongs to someone else?'. Correct answers, according to the testing manual, involve 'restitution or apology or both; mere confession is not satisfactory'. To the question 'What is the thing to do if another boy (girl, person) hits you without meaning to?' the 'only satisfactory responses are those which suggest excusing or overlooking the act' (for example, 'Tell them they never meant to do it'. Incorrect is, for example, 'I would hit them back').

In general, IQ scores correlate highly with scholastic achievements and their predictive value in this area is one use to which they have been put. Because scholastic achievement plays a large part in the choice of occupation, IQ scores, inevitably, have a reasonable correlation with socio-economic status. The question of whether high 'biological' intelligence and hence IQ determines scholastic success and hence (though less markedly) socio-economic status, or whether IQ tests merely test for the same things as do school examinations and give points for those responses that are likely to be associated with socio-economic success, is, of course, left unresolved by this type of correlation.

The point is that, far from being a relatively fixed measure, like height, for instance, the IQ test is a social construct. The normal curve of distribution of IQ scores does not arise 'naturally' from the data, but is produced by the deliberate choice of test items—those that do not discriminate adequately, or give skewed distributions, are deleted (they are often referred to as being inadequately 'g-loaded' items). When a new test is produced, it is 'standardized' against the old ones, and if it does not give comparable results to, say, the Stanford–Binet, it is discarded. To give an example of the consequences of this standardization procedure—in early versions of the tests, males and females scored differently on certain items. This result was felt to be inappropriate, and the tests were arbitrarily modified in such a way as to ensure that on the revised tests (within the White population) the sexes now score more or less identically.

Using the tests in Britain or the USA, it is generally found that working-class children score lower than middle-class children and rural lower than urban. In the USA, Blacks score lower than Whites (the mean for the Black population is about 85, or 1 standard deviation below the White level). Irish children in Ireland score lower than British children in Britain. IQ test scores may correlate with educational or socio-economic success, but our concern here is not this but whether they tell one anything about underlying biological, and hence possibly genetic differences, between individuals or groups, rather than about obvious social differences.

One problem is that even the 'culture-free' or 'culture-fair' IQ tests cannot adequately compensate for the known effects of group differences in what children learn to perceive as important, both between geographically separated cultures (advanced urban industrial versus rural peasant) or between classes*. An example is the study

---

* Anyone who has tried to do an OU CMA will be aware of the ambiguities that occur when different people attempt to read meaning into particular statements or questions. What is an 'obvious' meaning to the examiner is not necessarily so to the students, and students from widely different backgrounds may perceive differently just what is 'important' or 'obvious'!

made by Lewis, who argued that working-class children had to live in an environ-ment in which they are subject to much more misinformation ('noise') than middle-class children. He devised tests in which the person tested had to devise a strategy despite a great deal of such misinformation, and compared a group of working-class youths with a 'high-IQ' group. The working-class group did considerably better.

ITQ 25 Can one conclude from such a study that working-class children really have a higher intelligence than middle-class children?

Yet there is a reluctance on the part of many testers to accept the conclusion of ITQ 25. If the results come out in an 'unexpected' direction, alternative modes of explanation are sought. For example, there are certain items on which, in the United States tests, Black children score particularly well—Jensen and others arbitrarily define these as testing 'lower order', more routine, skills—the items on which the Whites score better are seen as indices of more 'creative' skills.

In fact, when gross IQ differences are pursued in more detail, they tend to evaporate or be capable of quite different explanations. For example, the observation that Irish children in Ireland score lower than British children, which has been used as evidence for a 'genetic' hypothesis, may be quite well accounted for by the fact that the children being studied were also being trained to be bilingual in English and Irish, a factor that is known to have at least a temporary retarding effect on school performance.

So far as Black/White differences are concerned, attempts to avoid the 'cultural' problem have included, for instance, doing IQ tests on Black and White groups in the USA matched for what is called 'socio-economic status', that is, groups doing roughly the same type of job, of the same age and educational background, etc. The group differences in IQ persist although the magnitude of the differences diminishes.

ITQ 26 Do you think this control is satisfactory?

It is interesting to note that in an ingenious experimental model working with several different strains of rat, Harrington (1975) was able to show that 'intelligence tests' —in this instance a variety of specially constructed tests of learning ability on mazes —*always* tended to favour the majority strain in any mixed population. In these experiments the tests were constructed and standardized separately for each of several strains of rat. Then mixed populations of the several strains were made, in each of which one strain formed the majority, the others the minority, and the tests were applied. On the tests standardized for the 'majority' strain, the minorities in the population always fared worst. Thus there appears to be an inherent tendency for tests that are standardized for particular populations automatically to favour the majority and disfavour the minority in this type of experimental design, simply as an artefact of test construction.

Yet another problem arises in the use of IQ tests, however, which springs from the actual test situation itself. Tests do not represent the application of a neutral instrument, a test, by an objective tester, to a person whose performance is being measured. Rather the results of the test are themselves the products of three-way interaction between the tester, test and the person being tested. Although the contribution of the person being tested to this product is probably the most sub-stantial, the other components of the interaction cannot be ignored. Such inter-actions manifest themselves, for example, at the level of labelling theory, in which the expectations of the teacher about the performance of the child modify that perform-ance (so-called Pygmalian effects, which are well documented, although still not universally accepted), or in reports that Black children may score better on IQ tests administered by a Black (or even by a computer!) than by a White.

Despite this, some educational psychologists continue to regard the tests, not just as one possible way of assessing an individual in a given social context and in relation to particular expectations and socially approved behavioural patterns— a possible useful clinical tool—but rather as a way of telling one something bio-logically real. We would suggest that the obsession with g is at best an index of the belief, that if something can be mathematically expressed and manipulated, then it is, by that criterion alone, science—part of the reductionist belief that a statistical phenomenon implies a genetic mechanism. g is a property that emerges from multifactorial statistical analysis, and the tendency to reify it—to regard it as *therefore* a property of the organism to which heritability estimates can be straightforwardly applied—is unhelpful.

15.3.3 The within-population heritability of IQ

Despite the difficulties inherent in the concept of IQ as a biologically based phenotype whose heritable component can be estimated, there have been, over the history of the mental-testing movement, a large number of efforts to apply the biometrical techniques of quantitative population genetics to the question of IQ variance. For the purposes of the discussion, we shall ignore the criticism of the previous Section and assume, for the sake of argument (as is done by the advocates of the claim that intelligence is largely inherited), that IQ is a measure of g and that g represents a meaningful behavioural phenotype. We can then review the data on the within-population heritability of IQ.

First, however, let us clarify what it is we are talking about.

QUESTION From your earlier work on this Course, and assuming IQ to be a measurable phenotype, how would you approach the answer to the question, 'how much does genotype, and how much does environment, contribute to the measured IQ score of an individual?'

ANSWER You should have recognized that this is a question to which *no* answer can be provided because we cannot measure either in any given individual; the phenotype is the developmental result of the expression of a particular genotype in a given environment, and an altered environment can change all aspects of this relationship.

We cannot talk about 'high IQ' or 'low IQ' genes, except in the context of a given environment, as you will recall from the PKU example (Section 14.3.4). (Incidentally, this makes biologically as well as sociologically fatuous the suggestion by one educational psychologist that the Black population in the USA has 'low IQ genes' because African chiefs sold off their more stupid people to be slaves or because the stupid ones ran more slowly and were therefore caught, and that the Irish population in Ireland has 'low IQ genes' because a preponderance of those with 'high IQ genes' emigrated to Britain or the USA.)

QUESTION Rephrase the previous question in a form that is potentially amenable to genetic analysis.

ANSWER The question should take the form, 'what proportion of the variance in IQ among individuals in a population is genetic, and what proportion is environmental?'

You should recall from Units 11 and 12 and from Section 14.2, that there are a number of approaches available to population geneticists in attempting to answer this type of question.

QUESTION What are some of the approaches that might be applied to a human population?

ANSWER As experimental manipulation is not possible, we are essentially reduced to approaches of the kinship-correlation type. That is, we estimate the similarity in IQ scores among individuals, ranging from MZ twins to unrelated individuals. From these data, we should be able to calculate h_B^2.

kinship correlations

QUESTION What is the biggest problem in such an approach?

ANSWER The problem of the environment. The more closely related individuals are, the more likely it is that they will live together and hence the more similar their environment will tend to be. Thus, the estimates will be impossibly confounded.

QUESTION Can you suggest a way round this difficulty?

ANSWER Look for related individuals who have been separated and reared in different environments. The 'cleanest' 'natural experiments' would be to examine the IQs of MZ twins who have been reared apart since birth. In such cases the *only* contributory factor to IQ differences must be the postnatal environment.

In what follows we shall look first at the MZ twin studies, and then other kinship correlation studies.

15.3.4 Separated twins reared apart

The number of identical twins who, once born, are separated shortly afterwards and then subsequently become available for study by educational psychologists, is not large, and up till 1974, there had only been four such studies. Table 16 (Kamin, 1975) gives the results—at first sight they look very convincing.

Table 16   IQ correlation studies in separated twins reared apart

	Study	Correlation between IQ scores	Number of twin pairs
1	Burt	0.86	53
2	Shields	0.77	37
3	Newman *et al.*	0.67	19
4	Juel–Nielsen	0.62	12

Note   Remember that a correlation of zero would imply no effect of genotype on IQ score; a correlation of 1.00, would imply no effect of environment.

These studies have formed the cornerstone of the argument that some 80 per cent of variance in IQ between individuals within a population is inherited. However, they have recently been subjected to extensive analysis by Kamin, and serious doubt has been cast on their validity. The biggest series is Burt's, and it was collected over many years. However, it appears that in many cases Burt did not apply standard IQ tests to the twin pairs, but gave them 'individual tests' or 'evaluated assessments', sometimes adjusted on the basis of teachers' reports, and sometimes in other, only partially reported, ways. The original data no longer exists, and the various reports that relate to the twins, over the 20 or so years in which they appeared, contain so many inconsistencies and have had such serious doubts cast on their statistical treatment that most authors (including Jensen) now discount the Burt data. (For the detailed discussion of the evidence, see Kamin, 1975.)

The second study is that of Shields, reported in 1962. The most important criticism here concerns what is meant by 'separated twins' in this study. The detailed accounts show that in only ten of the cases had members of the twin pair not attended the same school or been reared by related families—and even among this ten, six twins were given to family friends to be reared and in at least three cases lived in the same town, meeting one another in their homes. For the others in the study, the relationship was even closer: '. . . the paternal aunts decided to take 1 twin each and they have brought them up amicably living next door to one another in the same Midlands colliery village . . . they are constantly in and out of each other's homes . . .', or '. . . were not separated until their mother's death, when they were $9\frac{1}{2}$ . . .', are examples of what is defined as 'rearing apart'.

The other two studies are of very small numbers and have been subject to criticism concerning the nature of the statistical methods and standardization procedures used, which we shall not discuss here. Although there is ongoing controversy on whether any useful information can be salvaged from the twin studies, at present there are no reliable conclusions about the heritability of IQ that can be drawn from the MZ twin studies.

15.3.5   Other kinship correlations

Because of the difficulties of drawing conclusions from the separated twin studies, attention has been directed at other kinship correlations. The types of correlations that can be looked for are a comparison of MZ and DZ twins, sibs, parent–child relationships and studies of adopted children. There have been many studies along these lines, and some of the data then available were collected in a review by Erlenmeyer-Kimling and Jarvik in 1963 (Fig. 16). As in the separated-twin studies, these results too seem on the surface to imply a strong heritable component in IQ, and form one of the main planks of the 'hereditarian' thesis.

Again, we are indebted to Kamin for an exhaustive re-evaluation of the data, and without repeating all of the arguments here, we may summarize the conclusions.

First, in order to produce this apparent orderliness, many studies that give conflicting results are omitted, and arbitrary assumptions are made about those that

664

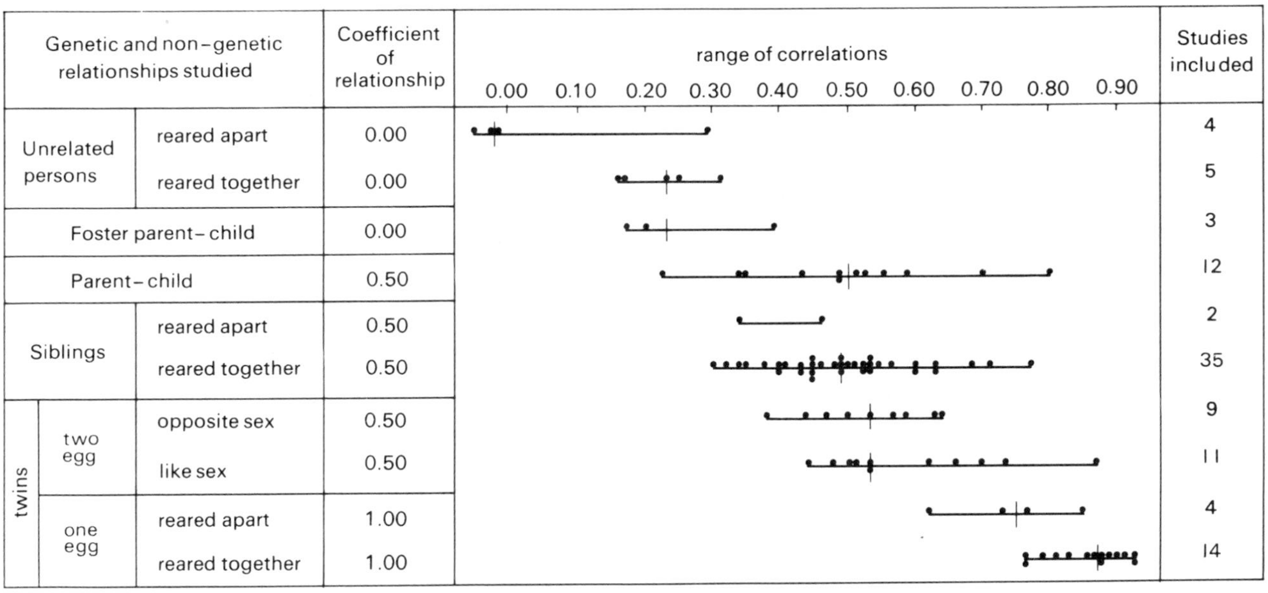

Genetic and non-genetic relationships studied		Coefficient of relationship	range of correlations	Studies included
			0.00 0.10 0.20 0.30 0.40 0.50 0.60 0.70 0.80 0.90	
Unrelated persons	reared apart	0.00		4
	reared together	0.00		5
Foster parent–child		0.00		3
Parent–child		0.50		12
Siblings	reared apart	0.50		2
	reared together	0.50		35
twins	two egg	opposite sex	0.50		9
		like sex	0.50		11
	one egg	reared apart	1.00		4
		reared together	1.00		14

are chosen. Second, and more important, the effects of environment and genotype interpenetrate in ways that are often not considered. We have already discussed the problem of age. A similar type of problem reappears in the comparison of MZ and DZ twins.

On genetic grounds MZ twins ought to be more alike in their IQ than DZs because they share a common genotype. In the simplest genetic models, a difference in IQ correlation between MZ and DZ twins of 0.3 implies a heritability (h_B^2) of 0.6. However, Schwartz and Schwartz (1974) showed that differences in correlation of about 0.3 are observed not merely between MZ and DZ twins, but between DZ twins of like sex compared to DZ twins of opposite sex, between male and female twins, and between the results of the separate studies shown in Figure 16. Further, the effect of environment is confounded with that of genotype because identical (MZ) twins tend to be treated more similarly than non-identical twins. Re-evaluating the data, based on a study of twins in the Philadelphia school system made by Scarr-Salapatek, the Schwartzes concluded that a maximum upper limit to heritability in the sample was $h_B^2 \leqslant 15\% \pm 16\%$. Once again, the empirical data are uncertain and give rise to a morass of conflicting interpretations supported by rival statistical procedures. They provide no foundation on which to mount any sort of theory from which a figure for heritability can safely be derived.

Figure 16 Kinship correlation coefficients for IQ (after Erlenmeyer-Kimling and Jarvik, 1963).

15.3.6 Theoretical considerations on the validity and meaning of heritability estimates

So far, we have presented the data on the kinship correlations and made statements about consequent heritability estimates, as if this did not in itself present problems. Yet you will know from Units 11 to 13 that to derive a heritability estimate, and to interpret it, requires not merely the plugging in of data to an equation, but some important genetic assumptions. We have discussed the nature and validity of these assumptions for selected cases of plant and animal breeding in Unit 13, but just how relevant are they to the genetics of human populations, and especially to phenotypes such as IQ? To put it straightforwardly, if the estimates of the broad-sense heritability of IQ in White populations in Britain and the USA were *really* around 0.8, as the hereditarian argument maintains (that is, discounting the counter-arguments advanced in the previous Section), what would it mean in relation to differences among individuals within the White population or differences between Whites and Blacks? Does it imply, as is generally interpreted, that 80 per cent of the difference between individuals or between groups in intelligence is inherited?

To answer this question, we have to consider the nature and meaning of the heritability estimate, which we introduced in Unit 12. Perhaps the best discussion of these topics in recent years is a paper written by Feldman and Lewontin (1975). Extracts of that paper are reproduced in Appendix 1. Here, we merely summarize the crucial points to the argument.

You should now read Appendix 1, which will also remind you what heritability means and how, in general terms, a heritability estimate is derived.

A heritability estimate, of either h_N^2 or h_B^2, is an algebraic expression which was originally developed in plant and animal breeding studies as a way of partitioning the variance of a character between environmental and genetic causes and another component representing genotype–environment interaction. However, the algebra on which it is based is an approximation that depends on the assumption that the range of environments considered is small and randomized across genotypes—a condition which is assured by the controlled condition of plant and animal breeding for which the algebra was developed. A heritability estimate is not a constant, and if the environment changes, so does the heritability estimate. Therefore the measure cannot predict the consequences of larger, and non-randomized, environmental fluctuations, and is essentially irrelevant to human population genetics, in which such fluctuations between environments within a particular generation and between different generations over a period of time are an essential part of the human condition. In a series of diagrams, Feldman and Lewontin (1975) show how hypothetical phenotypes vary according to their environment, and demonstrate that the heritability statistic does not allow any prediction about the effects of changed environments either on the phenotype or on the heritability statistic itself. They show the impossibility of distinguishing, in such cases, between what they call 'biological' and 'cultural' inheritance.

Finally, they make a criticism that has been frequently levelled at the use of heritability estimates in relation to group or 'race' differences, that is, the heritability estimate is a *within-population* statistic—it refers only to the proportion of variance that is genetic *within* a population, and can by definition say nothing about the differences *between* populations. You should know this from Unit 12, and be able to see that it would be possible to have two populations in which all the variance within each was entirely genetic and yet all the variance between the two was entirely environmental. There is no known algebraic or experimental method in genetics whereby one can derive an understanding of the causes of differences between two populations by a study of the causes of variance within each population.

Thus the heritability estimate is irrelevant to our understanding of the genetic and environmental causes of differences between human individuals, groups or populations, and can have no predictive value in either scientific or human affairs.

differences within populations

differences between populations

15.3.7 Can biology say anything about differences in intelligence?

Thus one may conclude that just as IQ measures can tell us little if anything about underlying biological mechanisms, so the apparatus of quantitative population genetics is a tool that cannot usefully be applied to an understanding of what contributes to the differences among individuals' performances on IQ tests. This is not, of course, to say that biology has nothing to contribute to the study of human learning or performance, or to adopt the 'straw-man' 'environmentalist' position, which is sometimes regarded as the opposite of 'hereditarian'. Both genetics and developmental neurobiology have contributions to make to our understanding of humans—but not if attempts are made to use inappropriate tools to ask fundamentally unanswerable questions. There is, to put it plainly, *no type of scientific research programme that could be devised to answer the question, 'how much do genes and how much does environment contribute to differences in IQ scores between different racial groups?'*. The question is not merely fallacious; it is, although *apparently* scientific, strictly meaningless: no amount of data collection can answer a question of this type.

To explore those biological factors that can be investigated and are known to be involved in human performance as expressed, for instance, in the result of the IQ test, would take us far outside the area of a genetics Course*.

For the present, we may merely note that among the predictive factors for a child's subsequent performance at school are included family size, socio-economic status, the mother's health during pregnancy and the child's birth weight. When these are improved, then school performance and 'IQ' improve in a manner that a naïvely 'hereditarian' analysis would consider extremely improbable. It seems unlikely that genetics will have much to contribute in the near future to speed this development, and indeed some of the 'pseudo-genetics' that has characterized the IQ debate may actually retard it. It is this observation that raises the last of the topics we wish to consider before leaving the 'race–IQ' debate.

* Those of you who are interested might like to follow another second-level Course, *The Biological Bases of Behaviour*, which explores these questions in more detail. (The Open University (1972) SDT 286 *The Biological Bases of Behaviour*, The Open University Press.)

15.3.8   Race and IQ: the history of an idea 1869–1969

To complete our discussion of race and IQ, we want to try to answer the question of why the issue has become one of popular concern over the last few years, whereas it was not, for instance, during the 1950s or early 1960s. One answer would be that there was now new evidence on the subject; however, this is not so—the kinship and twin studies have all been available for many years; the studies summarized in Figure 16 stretch back to the early part of the century, and Burt's hereditarian views spanned his long and active life. The evidence when Jensen's paper was written, then, was little more than what had been available when the UNESCO report reached a conclusion the reverse of his in 1951. To understand why old issues have re-emerged we have to look at the history of the mental-testing experiment and its links with eugenics, drawing for this purpose on the discussion by Rose (1975). We have already noted (Section 15.3.2) that testing originated with Galton and Binet. In *Hereditary Genius*, Galton (1869) was already convinced that there was a racial element in intelligence. 'The natural ability of which this book mainly treats is such as a modern European possesses in a much greater average share than men of the lower races.' For Galton, as for many of his successors, the purpose of the tests was to prove scientifically what they already knew in their bones to be true, that the British (or Anglo-Saxons) were a superior race, and the upper-class Englishman was biologically superior to the working-class Englishman (the question of 'social Darwinism' is dealt with in *HIST*, Section H.2.4).

When the testing movement crossed to the United States in the early twentieth century, it was developed by men who, unlike Binet, were convinced hereditarians and eugenicists. Only rarely was the tone as extreme as that used by the eugenicist Holland (1883), who claimed of the Blacks in the United States that 'Galton's law is squarely across their path, and the sooner they die gently out the better, and to assist them to multiply becomes as wrong as keeping the filthy and effete Turk in Europe for the sake of containing Russia . . .'.

None the less, Goddard, who was one of the pioneers of the hereditary view of IQ, argued that there was a close correlation between social-class position and the level of intelligence. He went on to conclude that those at the top of the pyramid had a particular responsibility to care for the masses in terms of 'social welfare' and 'national efficiency', and that in the context of a democracy it was important that the intellectual elite persuaded the masses to submit to their leadership. For Goddard, the virtue of the IQ test was that it demonstrated that he lived in a meritocracy, in which each citizen stood in the estate to which biology had been pleased to call him.

Nor was Terman (1924), who introduced the Stanford–Binet test, ideologically at odds with his predecessors; his eugenic concerns were supported by his belief that the social system was founded on the distribution of IQ.

> The racial stocks most prolific of gifted children are those from northern and western Europe and the Jewish. The least prolific are the Mediterranean races, the Mexicans and the Negroes. The fecundity of the family stocks from which our most gifted children come appears to be definitely on the wane. This is an example of the differential birthrate which is rapidly becoming evident in all civilized countries. It has been figured that if the present differential birthrate continues, 1 000 Harvard graduates will, at the end of 200 years, have but 56 descendants, while in the same period 1 000 South Italians will have multiplied to 100 000.

Other Americans pre-eminent in the testing movement, psychologists such as Yerkes, Hall, Brigham and Thorndike, were equally unequivocal in their conviction that Blacks were genetically inferior to Whites. As Thorndike (1927) put it, 'In the actual race of life, which is not to get ahead, but to get ahead of somebody, the chief determining factor is heredity'.

What is interesting is how, despite the distance in time and the advances in human genetics, we are now asked to return to the intellectual preoccupations of the eugenicists and mental testers of a past age. When 100 years after the publication of *Hereditary Genius*, we are asked to reconsider the Nature/Nurture debate, it must surely be clear that we are not dealing with a scientific question capable of scientific resolution, but one that primarily reflects political and social concerns.

It is not the job of these Units to explore the sociological reasons associated with the impact of the Jensen paper of 1969. It will be enough here to note that Jensen's own interest was not initially academic, but was concerned with the failure of an educational and social programme, Project Head-Start, itself a political response to what has been called 'the crisis of the cities' in the United States. At a time of rising general social and racial tensions—a situation that has characterized both the United States and Britain since the late 1960s—we should not be surprised to see genetics called in to 'support' the social structure, just as it was in the days of Galton or of Terman. Whether this a valid explanation of the flurry of interest in the IQ question, you may judge for yourselves.

15.4 Genetic change in present-day human populations

Up to this point we have examined the methods employed in studying human genetics, discovered the extent of variability in different human populations and seen the application of methods of population genetics in building up an account of human genetic history. The case study of IQ and its heritability illustrated the misconceptions and the difficulties as well as the varied ideological responses to the question of heritable components in complex phenotypes judged as socially important. There remains the future; does genetics have anying to say about the long-term prospects for humanity? After all, we stressed the predictive nature of genetic analysis at the start of this Course. In the Introduction to these Units we quoted H. J. Muller, a Nobel Laureate in genetics, describing his early vision of the promise that genetics offered to human society.

It seems logical to examine this question along the following lines. The human gene pool contains considerable genetic variation both manifest phenotypically among individuals in populations and hidden as recessive mutations in heterozygotes. Is the present-day composition of the human gene pool now stable or is it changing, or might it change in the future? The answer, which constitutes the remaining Section of these Units, comes from examining in turn the factors that might effect change in the human gene pool. The factors we shall consider are:

1 Natural selection in human populations.

2 Mutation.

3 Changes in the 'breeding structure' of populations.

4 The impact of medicine, in particular the effect of genetic counselling advice given to families in connection with the risk of appearance of a severely deleterious genotype.

5 Deliberate and positive attempts to alter, in a predetermined way, the genetic composition of humans by selective reproduction—in other words, the practice of eugenics.

15.4.1 Natural selection in human populations

The consequence of sustained natural selection is evolutionary change, but the essential preconditions are that there is genetic diversity and that different genotypes have different genetic fitness.

natural selection in human populations

> QUESTION What would be a general approach that might answer the question about whether genetic change is occurring in the human population?
>
> ANSWER Either of the following ideas may have occurred to you.
>
> 1 Measurement of the change in the frequency of a particular allele over a long period.
>
> 2 Estimation of the relative fitnesses of different genotypes in the population, perhaps by investigating deviations from the expected segregation.

The first alternative is going to take a long time and will involve several generations of geneticists! In principle, this approach ought to be very informative, but there is a risk of introducing a bias because of sampling difficulties in large populations. Moreover, any change in the breeding structure of the population (to be discussed in Section 15.4.3), such as a change in gene flow and genetic admixture as a result of increased geographic mobility, might also upset such measurements. If the gene flow is between populations with different allele frequencies, then the allele frequency would be found to change.

Turning to the second approach, how would we estimate differences in fitness or detect significant genetic loss? Natural selection depends both on differential mortality and differential fertility. In Western Europe and North America the expectation that a live-born child has of reaching reproductive maturity is very high, as mortality rates are low in the first 20 years. Thus, in these societies differential mortality may be less important than differential fertility as a 'motor' for natural selection and hence biological (as opposed to cultural) evolution.

Many studies have been made of the correlation between particular genotypes and susceptibility to various diseases that might remove those genotypes more frequently from the population (see also Units 9 and 10, Section 9.3). Table 17 (Lerner, 1968) shows the incidence ratio of four diseases in different phenotypes of the ABO blood-group polymorphism.

Table 17   The relationship between susceptibility to disease and blood group

Disease	Blood group	Ratio of incidence
gastric ulcer	O/(A + B + AB)	1.19
stomach cancer	A/(O + B)	1.19
pernicious anaemia	A/O	1.26
duodenal ulcer	O/(A + B + AB)	1.38

For example, the incidence of duodenal ulcer is 1.38 times as common in blood-group O individuals as in individuals of the other phenotypes.

> **ITQ 27** We assume that these diseases may occasionally kill individuals, thus removing their genes from the population. From the data in Table 17 for stomach cancer, would you expect that the frequency of I^O and I^B alleles would increase at the expense of I^A alleles in subsequent generations? Consider carefully what effects there would be if this disease removed people from the population at an average age of 50.

If a genotype has already produced its progeny before selection occurs (post-reproductive selection) then there will be no change in the zygotic frequencies in the next generation. This emphasizes the fact that one must remember how the nature of the reproductive cycle may interact with the process of selection and how the essential feature we are looking at is the relative frequency of different alleles in subsequent generations. Selection may eliminate or reduce the frequency of zygotes of a given genotype as soon as they are formed at fertilization. In humans, this might be very difficult to detect, as if the effective pregnancy spontaneously aborted, the selection might pass unnoticed. Thus, quite strong selection against some genotypes in the population might be hidden at this stage. If individuals carrying one particular genotype persistently had a shorter generation time than others (that is, they married and conceived children at an earlier age), then over many generations that genotype would increase in frequency. Similarly, if people of a given genotype had a significantly larger average family than did those of other genotypes, and the chances of reproduction remained constant for individuals from sibships of smaller size, then that genotype again comes to represent a higher proportion of the genotypes in the next generation. On the other hand, as has just been argued, the frequency of a genotype that has a shorter life expectancy, but that never dies before reproduction is complete, does not change in the next generation.

Far from being outside the range of interest of the geneticist, changing environmental factors are all-important in genetic analysis and prediction, especially for human populations. For a change, let us summarize the conclusions from earlier parts of this Course and *then* take a look at their implications for human genetics.

We now know that:

1   The same genotype may produce very different phenotypes in different environments.

2   Phenotypic identity can never be taken to mean genetic identity (see phenocopies, for example in Unit 8, Section 8.6).

The consequences may be that: according to statement 1, the same genotype may come to have very different selective values in different environments, and according to statement 2, when the environment is changed a series of genotypes with originally similar phenotypes come to have very different phenotypes. Previously un-acknowledged genetic heterogeneity will then have come to light.

The gut enzyme lactase provides an example of statement 1. If the enzyme is absent or defective the affected individual cannot break down and metabolize the lactose in milk, and as a result suffers diarrhoea and general discomfort. This enzyme is present in both children and adults in the Caucasian population, but in the majority (some 90 per cent) of the population in Africa and China the enzyme is present only in children and not in adults. It is possible that the presence of widespread dairying in Europe and America and its absence in Africa and China may be a secondary consequence of this genetic difference.

It could be argued that the use of fermented milk in which the bacterium *Lactobacillus* has performed the function of the lactase system has obviated the 'problem' of the 'genetic deficiency'. This example is quoted not just as an historical speculation about a difference in allele frequency between populations, but as an instance where a genetic situation, lactase deficiency in adults, may have favoured a particular social change, the diet, rather than the reverse. Of course, the reverse argument could be made, that the mutation to 'continued lactase production' in Caucasian populations may also have favoured its dietary habits.

If the environment changes (statement 2), then we conclude that the relative fitness of different genotypes may vary. It may be impossible to predict how selection pressure may change, particularly if it is not clear precisely what environmental factors have changed and which genotypes will become optimum under the new conditions. Genetic differences have been discovered in the sensitivity of individuals to certain drugs and for the 'sensitive' genotypes the 'standard' dose would be dangerous. For instance, the $G6PD$ deficiency is associated with sensitivity to Primaquine. Hb^A/Hb^S heterozygotes are 'sicklers' and their erythrocytes have very impaired oxygen-carrying efficiency at low partial pressures of oxygen. This genotype may become at risk at high altitudes or when using closed-system breathing apparatus. These two examples are chosen to illustrate that genetic variation for such genes is of great significance to the individual concerned. They show the danger that arises from the concept of a 'normal' phenotype or a 'norm' of reaction that does not really take account of large genetically based differences in response or sensitivity. It is not suggested that there is a high mortality associated with the use of particular drugs or breathing apparatus that removes particular alleles from the population, but the examples were taken to illustrate the principle that a change in environmental conditions might lead to certain genotypes giving rise to very deleterious phenotypes, with the possibility that selection might operate to alter allele frequencies. The con-verse situation may also arise in which a genotype, formerly under strong selection, is removed from that category by an environmental change so that the fitness of the genotype is increased. There are now several instances in which timely medical or surgical action can remove a genotype from the severely deleterious class to a situation where it suffers little if any impairment. We shall return to this question later when we consider the impact of medical genetic practice and genetic counselling on the gene pool.

norm of reaction

15.4.2 Mutation in human populations

ITQ 28 Achondroplasic dwarfism is due to a mutation with dominant auto-somal effect. In a Danish survey it was found that 457 normal sibs of dwarfs had 582 children between them (1.25 children per family). The 108 affected, dwarf, sibs had only 27 children between them (0.25 children per family).

(a) Calculate the coefficient of selection against these dwarfs based on their reproductive capacity. (Refer back, if in doubt, to Units 9 and 10, Section 9.5.)

(b) Predict the trend in the frequency of dwarfism in subsequent generations.

(c) Suggest a reason why in this study the frequency of dwarfs in the Danish population was found to remain constant over a number of generations.

This exercise illustrates that the measurement of selective coefficients alone would only show that selection was occurring and not necessarily that any change in allele frequency was occurring. In this particular case, the number of achondroplasic dwarfs born in the Danish population over a period of time was recorded. One could presume that if a dwarf was born to normal parents then it represented a new mutation, and in fact it was possible to estimate the mutation frequency this way. Among 94 075 babies born in a Copenhagen hospital, there were ten achondroplasic dwarfs, but only two of them had a parent who suffered with the same genetic effect.

> **ITQ 29** From these data, calculate the mutation rate (Units 9 and 10) for achondroplasic dwarfism per million gametes, remembering that the mutation has a dominant effect.

This exercise illustrates two points, the second of which relates to the previous Section on selection in human populations.

1 It shows how the mutation rate for a dominant allele in human populations can be calculated quite simply.

2 It shows that merely estimating the selective coefficient of a genotype does not permit one to conclude automatically that the allele responsible is changing in frequency in the population. An equilibrium may be established as a result of the removal of the allele by selection and its generation by mutation.

Another aspect of mutation as an important factor in producing genetic change in human populations is the need to be aware of the consequences of environmental changes that put people at risk from new mutagenic sources. The hazard from radiation sources, whether they derive from radioactive elements or from sources like X-ray machines used for medical diagnosis, has become clear. A number of compounds have also been found to be potent mutagens, and have been used for this purpose in genetic research in experimental organisms; but the other side of this coin is that the same or similar compounds may unintentionally become hazards because they may occur in manufacturing processes or be produced in our environment as a result of reactions of new synthetic chemicals.

> **ITQ 30** Give a genetic situation in which equilibrium exists between selection and mutation for a recessive allele (see Units 9 and 10, Section 10.3), what would be the consequence for the allele frequency if the mutation rate were to increase appreciably?

15.4.3 The breeding structure of the population

The *breeding structure* of a population refers to the genotypes involved in the pattern of matings that is occurring. In the Introduction to the population genetics Units (Unit 9), we examined some genetic properties of a hypothetical population in which there was random mating; that is, the frequency of each particular mating was the product of the frequencies of the relevant genotypes in the population. In human populations there may be departures from strict random mating. Furthermore, one has to be more precise in distinguishing marriages from 'matings', as not all marriages necessarily lead to the production of children and not all matings are marriages! Changes in breeding structure may lead to changes in the relative proportions of various combinations of alleles; that is, genotypic frequencies may change. This, in turn, may mean that selection, which tends to remove a particular allele from the population, may become more or less effective depending on how the genotypic frequencies change. We shall examine the effects of the following on the genetic structure of the population:

(a) inbreeding,
(b) non-random mating and
(c) phenotypic assortative mating.

breeding structure of a population

Inbreeding is the mating of individuals who are related genetically by descent. In quantitative terms the degree of inbreeding is usually expressed as the probability that the two alleles of a gene in a diploid individual will be identical by descent from a common ancestor. From this definition it follows that marriages in very small isolated groups, which must have often occurred in geographically isolated small towns and villages in previous centuries before mobility increased, probably represent quite substantial inbreeding.

inbreeding

> **ITQ 31** If the degree of inbreeding decreases, will the proportion of homozygotes in the population fall or rise or remain unchanged?

The appearance of phenotypes of the homozygous recessive genotype, including deleterious ones, are to be expected as a consequence of inbreeding. These may result from matings between genetically related individuals, known as *consanguineous matings*, for instance, marriages of cousins. Historical records of such marriages indicate that the proportion of consanguineous matings has been falling gradually from several per cent in the nineteenth century in Europe to less than 0.25 per cent in the 1970s. Inbreeding has been decreasing, partly because people move about more before marriage, partly because of a relaxing of social pressures that had formerly encouraged cousin and other consanguineous marriages, and partly because of increases in population size. In itself, inbreeding does not necessarily lead to changes in the gene pool, but there are frequently changes as a result of the 'exposure' of a great number of recessive homozygotes; selection against recessives exposed by inbreeding will increase, and because inbreeding is a consequence of a small population size, drift often accompanies inbreeding.

Deviation from random mating can alter the relative frequency of genotypes. Consider a situation in which the height of adults is affected by a number of segregating genes, and in which the homozygotes exert the extreme effects in the genes.

> **ITQ 32** If human mating is assortative by genotype, that is, 'like genotype' mates with 'like genotype', what will happen to the phenotypic variability for height in the population? Will the variance of height increase, decrease or stay unchanged?

We conclude that positive assortative mating by genotype will maximize the phenotypic variation between individuals.

> **ITQ 33** Suppose positive assortative mating by phenotype replaced random mating for a character. What would you have to know about that phenotypic character before you could predict whether this form of mating would increase variation in that character?

You will note that in themselves inbreeding and departures from random mating can only affect genotypic frequencies, but if those genotypes have different fitnesses, then both processes may contribute towards genetic change by selection. These changes in breeding structure affect the genetic contribution to the phenotypic variance for any character in a population.

15.4.4 The impact of medical genetics on the human gene pool; genetic counselling

> Estimates vary of the number of children born in Britain each year suffering from severe genetic diseases. A common figure that is quoted is 2% of all births, which amounts to 16,000 children a year, but a more realistic and verifiable figure is probably 0.4 to 0.5% of births: 3,000 to 4,000 children a year. Many more children, however, have various genetic predispositions. Genetic diseases are also the cause of some 11% of deaths of children in Britain: 16,000 infants died in Britain in 1971, about 1,700 of them from genetic diseases.
>
> Jones and Bodmer, 1974

This clearly points to the importance of genetics in medicine. The medical objectives must be to alleviate or remove such suffering, to understand the genetic basis of the inheritance and development of genetic diseases, and to offer counselling advice to individuals and families who may be 'at risk' in situations in which genetic diseases may be likely to re-occur. As well as examining these issues in this Section, we shall briefly look at the impact of the intervention of medicine on the human gene pool.

Consider a hypothetical situation in which a child with an abnormality clearly recognized as a genetic defect, or 'genetic disease', has been born to a couple. As well as learning to cope with this unfortunate family situation, the parents will probably have two important questions in mind:

1 What is the probability that a second similarly affected child will be born to them if they wish to increase their family?

2 What is the chance that their affected child, if he or she grows up and is fertile and wishes to marry, will pass on the mutation to his/her children?

The facet of medical genetics concerned with this problem of providing information, support and perhaps advice is called *genetic counselling*.

genetic counselling

> **ITQ 34** What would your prediction be as to the probability of a recurrence of the same condition in a subsequent pregnancy for the following cases?
>
> (a) A phenylketonuric (PKU) child born to normal parents. (This defect is due to an autosomal recessive gene.)
>
> (b) An achondroplasic dwarf born to normal parents. (This defect is an autosomal dominant mutation with high penetrance.)
>
> (c) A child with brachydactyly born to a couple, one of whom already showed this effect. (Brachydactyly is also caused by a dominant mutation.)

On the face of it this might suggest that counselling advice ought to be fairly straight-forward. However, although the ITQ reminds you of the basic genetic principles involved, it fails to reveal the true complexity of genetic counselling, which to be useful, must be precise. The number of known human genetic defects now runs in the hundreds and there is no reason why this number should not increase as new, previously unseen, or unexplained, defects are recognized and catalogued as genetic defects. Furthermore, humans are not so different in many respects from the other organisms you have studied in this Course, and so one would expect to find different genetic defects with similar phenotypes, phenocopies of gene defects, and mutations that were not always fully penetrant and therefore invisible. Last, but by no means least, there will be phenotypes that are controlled by polygenic systems.

Radio programme 14 examines particular aspects of genetic counselling, and in these Units we shall merely describe a number of technical developments that aid the diagnostician and clinician in genetic counselling. The particular developments are:

1 The sampling and examination of cells sloughed off the developing fetus during early pregnancy, to try to identify abnormal fetuses or those of a particular genotype. This process, which involves obtaining a sample of amniotic fluid, is called *amniocentesis*.

amniocentesis

2 The use of metabolic and biochemical tests to identify adult heterozygotes of what are generally considered to be recessive mutations.

3 The investigation of the karyotype of individuals (you may wish to look at Unit 5, Section 5.1, again to refresh your mind on this).

4 *Dermatoglyphics*—the analysis of changes in ridge patterns on the skin of the fingers and on the soles of the feet, as a broad indicator of the presence of a chromosome abnormality. (The study of dermatoglyphs, especially the total ridge count, is also an excellent model for the inheritance of a quantitative trait in humans.)

dermatoglyphics

The difference between prenatal diagnosis by amniocentesis and genetic counselling is that the former can provide specific information on the presence or absence of certain disorders, for example, chromosomal abnormalities, biochemical errors, indirectly diagnosable X-linked defects and some defects of the central nervous system. By contrast, genetic counselling can only be probabilistic in nature, estimating the chance that the child of a given couple will prove to be abnormal or defective. However, there are a number of other limitations that need to be borne in mind. In the first place, although amniocentesis can provide accurate information about the presence of, for instance, a biochemical abnormality, it does not auto-matically follow that if the affected child is born and develops, it will show an abnormality at the level of either physical or behavioural phenotype. A large num-ber of individuals who had they been subject to screening at birth would have been diagnosed as phenylketonurics, nevertheless grow up, untreated, to be 'normal' adults. Thus, the predictive value of a diagnosis by amniocentesis is itself limited. A second factor that must be considered is that the procedure of amniocentesis itself carries some risk both to mother and to the fetus. Any argument that it be generally applied must bear in mind that to do so would subject a large population to a relatively small risk in order to have the chance of diagnosing a small number of individuals who are at high risk. (Both these arguments of course apply in some measure to all mass-screening procedures.)

Although the focus of genetic counselling is on the immediate family, it may turn out that others in the wider family are 'at risk' as possible 'carriers' or heterozygotes for a deleterious genetic condition. It has been estimated that for every individual who seeks counselling as the result of the appearance of a known genetic defect in his or her family, there may be several relatives also at risk, though with a slightly lower probability.

Consider a couple who have had a galactosaemic child born to them—a recessive homozygote. The brother of the mother of this child wishes to know if he carries a heterozygote for the same defect. For a limited number of 'recessive' mutations it is possible to identify the heterozygote as physiologically distinct from normal, wild-type, homozygotes. Figure 17 indicates how heterozygotes for the recessive condition, *galactosaemia*, can be distinguished from normal homozygotes by measurements of the activity of the enzyme galactose-1-phosphate uridyl transferase.

galactosaemia

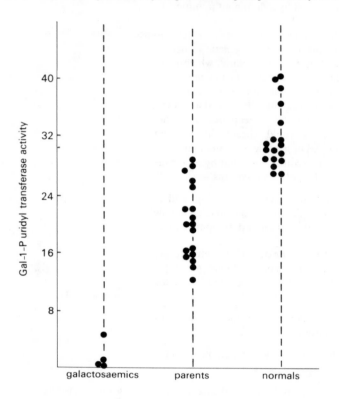

Figure 17 Uridyl transferase in 18 homozygous normals and 18 heterozygous parents of galactosaemic children and 4 homozygous affected children (after Stern, 1973).

The contribution of genetic counselling may be quite substantial, in terms of the alleviation of suffering and the prevention or reduction of stresses set up among other members of a family in which there is an individual with a severe genetic disease. However, the impact of the counselling on the genetic composition of the population in general may be rather different. It is untrue to say that all the consequences of genetic counselling are to 'improve' the human gene pool by ridding it of mutations with more or less severe deleterious effects, but it is equally incorrect to say that the intervention and efforts of medicine are sustaining deleterious alleles in the population. Consider the proposition that the contribution of medical genetics is to make available technological resources that render 'normal' genotypes that formerly had deleterious effects (for instance, by detecting and supporting phenylketonurics on defined diets so that they develop normally). Then one can say that the 'problems' associated with the segregation of this gene have largely been removed; in terms of population genetics the selective coefficient has been restored from a fractional value to nearly 1.0. On the other hand, if the consequence of counselling is that deleterious homozygotes of autosomal recessives are prevented from being born by selective abortion, then the remaining sibs are likely to include 'carrier' heterozygotes. If the family size in such a situation is, on average, identical to the family size of the general population, more heterozygotes would be produced and enter the population than if the same family size had included the severely deleterious homozygote which may not itself have reproduced. As a third point, it has been shown from 'follow-up' studies of couples given counselling advice when there is a risk of a severely disabled child, that many such couples have decided not to make further additions to their families.

As we have already mentioned, this topic is also treated in Radio programme 14, so we shall not expand this Section further. In fact, the discussion has strayed from the application of genetic counselling to individual families to the population in general, so this is a suitable point to move on to the rather vexed question of conscious intervention in human reproductive patterns for genetic 'benefits', which, as we hope you will realize, we are justified in putting in inverted commas.

15.4.5 Eugenics: selective breeding in humans

A programme of eugenics is a proposal to effect deliberate and chosen changes in the human gene pool so as to produce, as the goal, particular phenotypes, by fostering, encouraging or even coercing selective reproduction or selective breeding. In naïve terms, its counterpart may be seen in the use of genetic expertise in applied plant and animal breeding (Unit 13). There is no doubt whatsoever that this offshoot of genetics has led to controversy and has been with us in one guise or another ever since—and even before—the beginning of genetics.

eugenics

At this point it would be useful for you to read again Section H.4.1 on the eugenics movement in HIST.

The major part of this Section comprises a series of quotations from various authors, some of them geneticists, and notable ones at that, and others who are not geneticists. In part, they speak for themselves, but while you are reading them, we ask you to consider two things in particular:

1   How far do their conceptions reflect their view of society rather than the narrower internal logic of the science of genetics (*HIST*, Section H.0)?

2   How feasible are the views portrayed of the application of genetic methods, seen as a development of the internal logic of genetics?

These Units opened with a quotation from the autobiography of the geneticist H. J. Muller. Below is a brief introduction to the potential for genetic improvement that Muller envisaged, from *Out of the Night*, which was published in 1936.

> Make considerations of reproduction dictate the expression of personal love, and you not infrequently destroy the individual at his spiritual core; thus 'eugenic marriages' cannot as a whole be successful, so far as the parents are concerned. On the other hand, make personal love master over reproduction, under conditions of civilisation, and you degrade the germ plasm of the future generations. Compromise between these two policies, and you cripple both spirit and germ. There is only one solution—unyoke the two, sunder the fetters that from time immemorial have made them so nearly inseparable, and let each go its own best way, fulfilling its already distinct function. The physical means for this emancipation are now known for the first time in history.
>
> When we consider what the recognition of this principle would mean for the children —for those future men and women the character of whose lives we cannot in any event escape the responsibility of predetermining—our obligation becomes clear and compelling. The child on the average stands half-way, in its inherited constitution, between its father and the average of the general population, and so it would be theoretically possible even now—were it not for the shackles upon human wills in our society—so to order our reproduction that a considerable part of the very next generation might average, in its hereditary physical and mental constitution, half-way between the average of the present population and that of our greatest living men of mind, body, or 'spirit' (as we choose). At the same time, it can be reckoned, the number of men and women of great, though not supreme, ability would thereby be increased several hundredfold. It is easy to show that in the course of a paltry century or two (paltry, considering the advance in question) it would be possible for the majority of the population to become of the innate quality of such men as Lenin, Newton, Leonardo, Pasteur, Beethoven, Omar Khayyam, Pushkin, Sun Yat Sen, Marx (I purposely mention men of different fields and races), or even to possess their varied faculties combined.
>
> We do not wish to imply that these men owed their greatness entirely to genetic causes, but certainly they must have stood exceptionally high genetically; and if, as now seems certain, we can in the future make the social and material environment favourable for the development of the latent powers of men in general, then, by securing for them the favourable genes at the same time, we should be able to raise virtually all mankind to or beyond levels heretofore attained only by the most remarkably gifted.

QUESTION What are the social and genetic assumptions that Muller has made in developing this proposal?

ANSWER Firstly, he has presumed that the qualities he associates with Lenin, Newton and others are phenotypes that would be most desirable in society, and that others would agree with his choice. Secondly, without providing any evidence in support, he has assumed that these peoples' phenotypes accurately reflect genotypes very different from 'commonplace' combinations of genes. Thirdly, there is the assumption that it will be possible to detect the 'correct' 'superior' genotypes in each generation. Fourthly, he assumes, rather like the German farmer of Unit 1, that 'innate quality' is an exclusively male characteristic, inherited, apparently, as will be seen in the next quotation, in a sex-linked manner with no female contribution other than that of being a 'receptacle'. Muller visualized that as an aid to assessing the 'genetic worth' of an individual his sperm might be stored for a period.

If this could be done, it might be a salutary rule, eventually, not to make use of such material in any considerable way until say, twenty-five years after the death of the donor. After this length of time society would often be in a far better position to judge fairly of a man's achievements and his general worth, than during the heat and struggle, and the possible intrigue and bias, of his own personal life; there would also, by then, have been some opportunity to judge of his genes by observations on the character of a limited number of his progeny. How fortunate we should be had such a method been in existence in time to have enabled us to secure living cultures of some of our departed great! How many women, in an enlightened community devoid of superstitious taboos and of sex slavery, would be eager and proud to bear and rear a child of Lenin or of Darwin! Is it not obvious that restraint, rather than compulsion, would be called for?

Published speculations about a 'eugenic utopia' are legion. The following extract from the American sociologist, Kingsley Davis (writing with tongue in cheek?) serves at one and the same time as example and commentary on itself.

A first step in the construction of a eugenic utopia is to decide whether the entire population, or merely an elite, is to have its heredity improved. If the elite idea is favored, then obviously the society cannot be democratic. Either the hereditary elite would rule by virtue of its superior birth (the antithesis of democracy) or it would be governed by the hereditary riff-raff (a paradoxical disproof of the controlled-hereditary idea).

A second step is easier: it is to imagine a social adjustment whereby those having their heredity controlled (whether an elite or the whole population) would have no preference for their children, in the sense of genetically their 'own'. This change of attitude could presumably be accomplished by educating couples to welcome a child which comes from artificial insemination or, better, from an implanted fertilized ovum. In the latter case the child would be genetically derived from a male and female who were superior to the couple, but it would be born to the 'mother', would be nurtured by her, and would consequently be emotionally identified with her. The parents would thus regard the child as their own—much as a purchased house or car becomes a source of pride to its new owners, regardless of the fact that they themselves did not manufacture it. The nation could maintain a board of geneticists to determine who should furnish the sperm and the ova and what crosses should be made in the artificial mating. Needless to say, the males not required to supply sperm to the official 'sperm banks' would all be sterilized, and the women not supplying ova would have their own ovulation either suppressed or diverted. The board of supervising geneticists would have confidential records on the pedigree of all persons born in the population, as well as records of their traits and achievements. . .

Although modern research is giving a greater role to maternal love and attention in personality development than was formerly recognised the fact remains that professional child-rearers—trained, and perhaps even bred, for the task—might be superior. Once this happened, there would be little reason for encouraging stable marriage and family life. A sizable proportion of women would be required to serve as maternal host for bearing a number of offspring. Other women would be freed from this duty but might nevertheless be professional child-rearers. The sterilized men would be free to enter any kind of relations that suited them. The sperm- and ova-bearers, however, would have to be carefully regulated, perhaps allowed to mate amongst themselves, but not necessarily allowed to rear children. Obviously, under these conditions, marriage and the family as we know them might cease altogether. With sexual behavior divorced from reproduction, why regulate it at all? With reproduction divorced from child-rearing, why build up an identity between two 'parents' (male and female) and their 'offspring'? If social identity is necessary for children, they can be emotionally attached to one or more professional child-rearers. Presumably the relations among any set of child-rearers would not be complicated by sexual possessiveness, preoccupation with pregnancy, the necessity of coping with children of disparate ages, etc. The business of socialization, like the business of genetic selection, could thus be rationalized along scientific lines, utilizing an intelligent division of labor.

Quoted in Roslansky, 1966

Most discussions concerning genetic interventions end up in the area of politics and ethics. As a prelude to a discussion on the ethical issues, the American sociologist A. Etzioni (1975), in his book *Genetic Fix: The Next Technological Revolution*, has categorized all possible genetic interventions in humans in the following way:

First it seemed helpful to separate genetic interventions used for *therapeutic purposes* (e.g. to curb sickle-cell anaemia) from those used for *breeding* purposes (that is, to 'order' a child with certain desired attributes (e.g. six feet tall and with red hair), the way attributes of racehorses and show dogs can be specified in advance).

Next it seemed useful to distinguish between genetic interventions introduced to *serve individuals* (e.g. parents who wish a normal child or a child of high IQ) and those used to promote *societal*, or public, policy (e.g. stamp out disease, breed wiser people).

By crossing the two dimensions the way we cross coordinates to locate places on a map, it seemed possible to locate the various issues.

I realized later that I had to make a distinction concerning the *method* of intervention used. And so I further divided the societal section into *voluntary* controls, (e.g. the way we ask, but do not force, people to limit their family size) versus *coercive* controls (e.g. the way we [in the U.S.A] make couples take a Wassermann test to rule out syphilis before they marry).

The final chart looked like this:

	Therapeutic Goals	Breeding Goals
Individual Service	1. e.g. abort deformed fetuses on demand	3. e.g. artificial insemination; parents' choice of donors' features
Societal Service *Voluntary*	2. e.g. encourage people to abort a deformed fetus	4. e.g. urge people to use sperm from donors who have high IQs.
Coercive	e.g. require a genetic text before marriage license is issued	e.g. prohibit feeble-minded persons from marrying

... Thus the issue of whether or not a mother should be free to abort a deformed fetus belonged in cell 1, together with other individual therapeutic questions. The issue of whether or not society should promote genetic tests and abortions to curb genetic illnesses belonged, together with other public health issues, in cell 2. The question of individuals having the right to design their next child, a quite different issue from the therapeutic one, found a place in cell 3. If society could follow a policy leading to a 'better' human stock, that belonged in cell 4.

You might consider Etzioni's table—especially in the light of the radio discussions of genetic counselling, *genetic engineering* and eugenics—and decide how useful it might be in reality? For instance, where would you place research on the hazards of environmental mutagens and the recommendations for safety procedures that derive from this research? Do you also consider the distinction between 'therapeutic' and 'breeding' goals to be a helpful one? And at a more general level, just who are the 'experts' who are to make the decisions?

genetic engineering

Having become attuned to the language of genetics during this Course, you will doubtless come across many writings similar to those just given. The quotations we have included did not qualify by any intrinsic merit or reflect any value judgements of our own but rather indicate the wide spectrum and the many levels at which discussion of humanity's genetic future is being held.

At this point in our discussion, the interaction between the 'internal logic' of the development of genetics with social, political and ethical questions of the widest importance to the sort of people we hope our children and our children's children will be, and the sort of society in which they live, must take us far beyond the bounds of this Course—but not, we believe beyond the bounds of any of us—Course Team and students alike, for thought and action.

Appendix 1   The heritability hang-up

These extracts are taken from 'The heritability hang-up' a paper by M. W. Feldman and R. C. Lewontin, published in *Science*, *N.Y.* **190** (1975), 1163–8.

... The analyses and arguments that were made by Jensen (1969) for IQ, and by others for other quantitative traits in humans, are all based on a fundamental methodology that was invented by R. A. Fisher, the analysis of variance. The analysis of variance is meant to cope with the problem of dissecting the multiple causes of observed phenomena when the actual physical chain of causation of each individual event cannot be followed. What is observed is the variation in the phenomenon, measured quantitatively by the variance: the analysis partitions this variation into a proportion that is ascribed to the variation in each causal element and each combination of causal elements. Thus for IQ, the total variance in IQ scores in a population would be partitioned into an environmentally caused variance due to variation in the life experience of individuals, a genetic variance arising from variation in heredity among individuals, a genotype–environment interaction variance reflecting the lack of additivity of genetic and environmental deviations, and an error variance arising from uncontrolled variations in test procedures and, more important, developmental accidents that cannot be associated with specific, known environmental variables. As we show below, this partitioning of the causes of variation is really illusory, and the analysis of variance cannot really separate variation that is a result of environmental fluctuation from variation that is a result of genetic segregation. The genetic variance depends on the distribution of environments and the environmental variance depends on the distribution of genotypes. The analysis of variance is, in fact, what is known in mathematics as a local perturbation analysis. It is assumed that the actual IQ of an individual is some unknown function of genotype (G) and environment (E)

$$IQ = f(G, E)$$

In any given population, there is some joint distribution of genotypes and environments

$$\phi(G, E)$$

and this joint distribution is mapped onto a distribution of IQ scores $\theta(IQ)$ by a functional equation.

A complete analysis of the causes of variation would involve predicting the changes in the IQ distribution $\theta(IQ)$ from changes in the distribution of genotypes and environments $\phi(G, E)$. However, such an analysis would require that we know the first partial derivatives of the unknown function $f(G, E)$. What we substitute instead is an analysis of what would happen to the mean of $\theta(IQ)$ if very small perturbations were made in the mean of $\phi(G, E)$. The analysis of variance is a way of estimating the effects of these very small perturbations in the means, and the variance components estimated are directly related to the partial derivatives of the unknown function $f(G, E)$. Thus the analysis of variance produces results that are applicable only to small perturbations around the current mean. It cannot make any predictions about any larger issues.

In the analysis of variance of genetic and environmental causation, a special term is used for the proportion of all variance that is partitioned into the genetic variance. This proportion is called heritability in the broad sense (h_B^2). The genetic variance itself can be further broken down into a contribution that is due to individual alleles (additive variance), a contribution that is due to pairs of homologous alleles at a locus (dominance variance), a contribution that is due to combinations of non-homologous loci (epistatic variance), and so forth. The proportion of the phenotypic variance that is additive genetic variance is called heritability in the narrow sense (h_N^2).

We claim that this type of formulation is irrelevant to human population genetics on two counts. First, a model that is structured in this way cannot produce information about causes of phenotypic differences. Second, we do not, nor can we, use variance analysis in the resolution of those problems that are acknowledged to be central to the study of human population genetics ...

Normal Human Variation: Quantitative Characters

Perhaps the best publicized, most controversial, and least understood area in which variance analysis has been applied is in the study of normal (nonpathological) quantitative phenotypic characters (that is, those that are not simply Mendelian but which have unknown or complex patterns of inheritance (or both)). We have already introduced the concepts of broad and narrow heritability as they are used on characters of agricultural importance and animal breeding. In particular, the heritability of an economically valuable trait is useful in the design of breeding programs whose goal is to change the distribution of the phenotypic measure in a restricted population under a precise set of environmental conditions. We now present an analysis of why the application of this type of analysis to human behavioural traits cannot help to clarify the causes of a phenotypic measure. Our arguments are especially pertinent to the IQ controversy.

We start with the assertion that, for the quantitative characters in which we are now interested, differences in both genotypes and environment can be causes of phenotypic differences. What is not so well accepted, however, is that analysis of variance and its summary statistic, heritability, do not separate the two causes of variation in the phenotypic measure. This is because the analysis of variance is done (and heritability is calculated) with respect to a particular array of genotypes and environments in a specific population at a specific time. This array is usually a biased sample of the full array of genotypes and environments ...

Figure 1 is a norm of reaction figure that gives the phenotype P as a function of the environment E for two different genotypes G_1 and G_2. Obviously, both genotype and environment influence the phenotype in this example. However, if the environments are symmetrically distributed around E_1 (Fig. 1), there will appear to be no average effect of genotype, while if the population is weighted

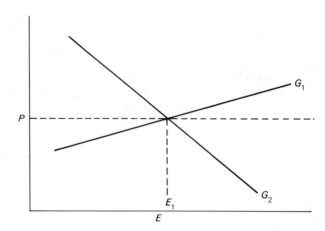

Figure 1 Phenotype P plotted against an environmental variable E. If the environments are symmetrically distributed around E_1 there is no average effect of genotype. If there is an excess of G_1 in the population the average phenotype will be constant, as represented by the horizontal dashed line.

toward an excess of G_1, the average phenotype across environments will be constant, as is shown by the dashed line. Thus the environmental variance depends on the genotypic distribution, and the genotypic variance depends on the environmental variance. This very important interdependence means that for a character like IQ, where the norm of reaction, the present genotypic distribution, and the present environmental distribution are not known, we cannot predict whether an environmental change will change the total variation ...

A further important point shown by Figure 1 is that fixing either the environment or the genotype does not necessarily lead to a decrease in the total variance. For example, fixing genotype G_2 (and thus eliminating the genetic variance) increases the total variance because G_2 is more susceptible to environmental change. It is also easy to construct graphs like Figure 1 in which environmental change improves the phenotype of both G_1 and G_2 but decreases the proportion of the variance that is genetic.

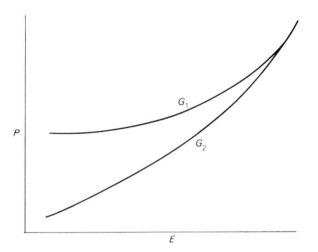

Figure 2 Phenotype P plotted against an environmental variable E. If the environments are weighted to the left there is a strong genotypic effect. If the environments are weighted to the right the genetic variation is reduced.

Figure 2 is a case such that, if the environments are weighted to the left, analysis of variance shows a strong genotypic effect. If the environments are weighted toward the right, thus producing improvement in both phenotypes, the proportion of variation that is genetic is reduced. . . [Further, the] cultural inheritance can have a pronounced effect on the phenotypic mean in a very short time. In fact, even in the absence of genetic variation, correlations between relatives may be expected from cultural effects alone. When both biological and cultural inheritance occur, separation of these effects is bound to be extremely difficult, especially in the absence of reliable data on adoptions . . . Our recent work indicates that cultural effects can strongly influence gene frequency changes as well as overcome the effects of strong natural selection in the sense that phenotypes acquired by learning or other modes of cultural transmission can spread through a population even though they lower the fitness of the individuals showing the phenotype. In the process, gene frequences also change. . .

Heritability and Differences between Groups

Jensen (1973) states that 'the fact of substantial heritability of IQ within populations does increase the *a priori* probability that the population difference is attributable to genetic factors'. Many authors have pointed out the logical flaws in this statement and counter-examples have been presented. For purposes of argument, consider the case of skin color. If we estimate the heritability of skin color among white New Yorkers, including people of Italian, English, Irish, Puerto Rican, and Polish ancestry, we would find

a high heritability. Suppose we now compare a group of New Yorkers who are left to winter in the city with a group of their well-to-do in-laws who spent the winter in Miami Beach. There would be a considerable skin color difference between groups, but no genetic causation.

... [In summary] we are unable to make any inferences from between-group differences and within-group statistics about the degree of genetic determination of the between-group differences. In other words, the concept of heritability is of no value for the study of differences in measures of human behavioural characters between groups.

Historical Reconstruction

To see a legitimate use of heritability analysis in human populations, we need to go back to Fisher's original ideas of genetic variance, and especially to his fundamental theorem of natural selection ... (1958). This theorem states that the rate of change of fitness is equal to the additive genetic variance in fitness. It is a consequence of the definition of additive genetic variance as the regression of offspring phenotype on parental phenotype that a parallel theorem holds for any phenotypic character. The change in mean phenotype is equal to the ratio of the additive genetic variance to the total variance (that is, h_N^2) multiplied by the selection differential.

It can also be shown that as selection progresses, the additive genetic variance is 'used up' so that the h_N^2 is decreased finally to zero, or nearly so. A consequence of these theorems is that, if natural selection has long been in operation on a character, the additive genetic variance for the character should be small, and the only genetic variance left should be nonadditive (dominance and epistatic variance). Thus we may be able to judge, from the ratio of h_N^2, which goes to zero during evolution, to h_B^2, which does not, how much selection has gone on. For example, if we really believed the estimate of 0.75 for the h_B^2 of IQ in European populations (we do not), and if we believed the single published estimate of h_N^2 of 0.40 (we do not), we would be forced to conclude that whatever it is that IQ measures, it has not been under intense selection for very long. Conversely, if there is a great deal of nonadditive genetic variance, but very little additive, we may guess at a long and consistent history of selection.

Of course, these are only weak inferences since, in the absence of knowledge about selection intensities, we cannot specify what we mean by long and intense selection. In addition, because of genotype–environment interactions, especially in behavioural traits, a long history of selection in one set of environments may reduce the h_N^2 to a very low value, but a recent change of environment may produce a new level of additive genetic variation.

Conclusions

The problem we have been examining is the degree to which statistical structures can reveal the underlying biological structure of causation in problems of human quantitative genetics. We must distinguish those problems which are by their nature numerical and statistical from those in which numerical manipulation is a mere methodology. Thus, the breeding structure of human populations, the intensities of natural selection, the correlations between mates, the correlations between genotypes and environments, are all by their nature statistical constructs and can be described and studied, in the end, only by statistical techniques. It is the numbers themselves that we need for understanding and prediction.

Conversely, relations between genotype, environment, and phenotype are at base mechanical questions of enzyme activity, protein synthesis, developmental movements, and paths of nerve conduction. We wish, both for the sake of understanding and prediction, to draw up the blueprints of this machinery and make tables of its operating characteristics with different inputs and in different milieus. For these problems, statistical descriptions, especially one-dimensional descriptions like heritability, can only be poor and, worse, misleading substitutes for pictures of the machinery. There is a vast loss of information in going from a complex machine to a few descriptive parameters. Therefore, there is immense indeterminacy in trying to infer the structure of the machine from those few descriptive variables, themselves subject to error. It is rather like trying to infer the structure of a clock by listening to it tick and watching the hands. At present, no statistical methodology exists that will enable us to predict the range of phenotypic possibilities that are inherent in any genotype, nor can any technique of statistical estimation provide a convincing argument for a genetic mechanism more complicated than one or two Mendelian loci with low and constant penetrance. Certainly the simple estimate of heritability, either in the broad or narrow sense, but most especially in the broad sense, is nearly equivalent to no information at all for any serious problem of human genetics.

Note The references we have retained in this extract are to:

Fisher, R. A., (1958) *The Genetical Theory of Natural Selection*, Dawes.

Jensen, A. R., (1969) *Harvard Educational Review*, **39**, 1–123.

Jensen, A. R., (1973) *Psychology Today*, **7**, 80.

Self-assessment questions

Section 14.1.3

SAQ 1 (a) What is the distinction between heterokaryons and hybrid cells? (b) State two uses for each in genetic experimentation.

Sections 15.1 and 15.2

SAQ 2 In humans, which of the following characters (1–7) is or has been of relevance to:

A   the social ascription of race

B   the biological ascription of race

C   neither

1   skin colour

2   native language

3   blood-group frequency

4   religious affiliation

5   enzyme polymorphism

6   frequency of occurrence of PKU

7   height

Section 15.3

SAQ 3 How is an IQ score measured?

Section 15.4

SAQ 4 Summarize the main reasons that we have given why IQ is not a convincing measure of 'intelligence'.

Sections 14.1, 14.3 and 15.3

SAQ 5 Assuming that it were possible to provide an objective assessment of the phenotype, which of the following types of experimental method (1–7) might be used to cast light on the possible mode of the genetic transmission of, and the heritability of:

(a) schizophrenia

(b) intelligence?

1   somatic-cell genetics

2   the study of MZ and DZ twins reared together

3   biochemical differences in blood proteins

4   the study of MZ twins reared apart

5   the comparison of properties of heterokaryons from schizophrenic and non-schizophrenic individuals, or low and high IQ individuals

6   kinship correlations

7   pedigrees

Section 15.3

SAQ 6 Under what conditions might one expect to be able to measure the heritability of differences in a character such as height between Black and White populations in Britain or the USA?

Section 15.4

SAQ 7 Summarize the arguments for and against selective breeding programmes in which sperm banks derived from 'great men' are established on the lines that Muller advocates.

Section 15.4 and general

SAQ 8 Consider the potential genetic interventions (1–5), and on the basis of your knowledge of the Course, including the TV and Radio programmes, classify them as technically

A   imminent

B   potentially likely to occur within the next decade

C   possible within the next 50 years

D   inherently improbable

1   the elimination of the PKU allele by breeding restrictions

2   the mother's choice of the sex of her next child

3   amniocentesis for diagnosis of potential genetic disorders

4   genetic therapy for disorders such as diabetes by replacement of the deficient gene(s) for insulin

5   selective breeding for people with high IQs

Answers to ITQs

ITQ 1 (*Objective 2*) Brachydactyly is caused by a dominant allele. It shows an unbroken line of transmission through the five generations. In each generation the ratio of brachydactylous to normal sibs (brothers and sisters) is not significantly different from a 1 : 1 ratio, and brachydactyly affects males as often as females. Thus, the gene in question is autosomal.

ITQ 2 (*Objective 2*) The phenotype of the affected individuals is due to an autosomal recessive allele. The heritable condition is known as phenylketonuria (PKU), an upset primarily related to a failure to metabolize the amino acid, phenylalanine, with consequent deleterious effects on brain development (see Section 14.4).

ITQ 3 (*Objective 2*) Marriage 1 produced four affected children (two males and two females) in a sibship, yet neither parent was a deaf-mute. We can suggest that the phenotype is due to an autosomal recessive allele. No segregation occurs in the progeny of marriage 2 (two males and two females), suggesting homozygosity for an autosomal recessive allele. In marriage 4 (of family 1 with family 2) all four children are affected. So we conclude that deaf-mutism in families 1 and 2 has the same *underlying* genetic defect. Intermarriage between families 3 and 4 produced no deaf-mutes in a sibship of six. We conclude that deaf-mutism in families 3 and 4 is due to different genetic factors. These marriages represent highly informative complementation tests!

ITQ 4 (*Objective 4*)

A 4

B 1

C 4

Statement A The only difference between MZ and DZ twin pairs is the genetic differences within the DZ pairs. All other differences are equalized between these groups. Therefore, we are forced to conclusion 4.

Statement B The members of a DZ twin pair are no more alike genetically than any two sibs. The fact that their intra-pair differences are similar suggests that the difference between the two groups (birth order) does not significantly increase the difference (conclusion 1).

Statement C We have already argued that the similarity between DZ twin pairs and sibs shows that birth-order differences are unimportant here, so we must conclude that it is genetic differences that are responsible (conclusion 4). Conclusion 3 is an incorrect statement and conclusion 2 would be upheld only if the MZ and DZ curves were superimposed and the curves for the sibs were distinct from them.

ITQ 5 (*Objective 4*) It indicates that there are genetic differences between the DZ twin pairs that affect their susceptibility to these pathological conditions. It is possible that you did not conclude this, as both these conditions are produced by a recognizable *environmental* factor (in the case of polio, a virus and for rickets, vitamin D deficiency). Note that the case of DZ twins is interesting because presumably the environments are likely to be the same and yet there is a marked discordant effect.

ITQ 6 (*Objective 3*) It has been possible to recognize phenotypic differences 1, 3, 4, 5, 6, 11 and 13 in cultured cells. That 1, 3 and 11 may be recognized should be apparent from the previous paragraph. Temperature-sensitive lethality (phenotype 4) is examined simply by shifting growing cells to higher temperatures. A pigment-formation difference (phenotype 5) will be an obvious visible change and an antigenic change (phenotype 6) can be recognized by the response of the cells to specific antisera (whether the cells bind the antibodies or not).

ITQ 7 (*Objective 3*) Statements 2 and 3 are correct, and 1 and 4 incorrect. To grow in HAT medium, a cell, hybrid or not, must have a functional TK gene. The mouse genome does not, so one of the three human chromosomes still present presumably carries TK (statement 2) and 'protects' this hybrid. A functional TK gene allows a cell to take up and incorporate nucleosides (Fig. 8). If it takes up bromodeoxyuridine (BUdR) the cell dies, so if it happends to *lose* the chromosome carrying the TK⁻ gene, the hybrid will be unable to use BUdR and will survive (statement 3). Differences at the HGPRT locus have not been examined, because we did not expose the cells to 8-azaguanine.

ITQ 8 (*Objective 6*) Table 18 is an expanded form of Table 8, including Fisher's predictions for antigens CDE and CdE, and antibodies anti-d and anti-e.

Table 18

| | Red-cell antigen phenotypes | | | | | | | |
Antibodies	CDe	cDE	cde	cDe	cdE	Cde	CDE	CdE
anti-C	+	−	−	−	−	+	+	+
anti-D	+	+	−	+	−	−	+	−
anti-E	−	+	−	−	+	−	+	+
anti-c	−	+	+	+	+	−	−	−
anti-d	−	−	+	−	+	+	−	+
anti-e	+	−	+	+	−	+	−	−

ITQ 9 (*Objective 6*)

cdE (1.2 per cent) is produced by a cross-over between *D* and *E*

Cde (1.0 per cent) is produced by a cross-over between *C* and *D*

CDE (0.2 per cent) is produced by a cross-over between *C* and *E*

These data suggest that crossing over between *D* and *E* is the commonest type, and hence that *D* and *E* are the furthest apart. The order is therefore *DCE* (or alternatively *ECD*).

ITQ 10 (*Objective 6*) The *dCE* antigen is very rare because it can be obtained only by a cross-over in the heterozygotes between *dCe*, *DCE* and *dcE*, and these are the least common genotypes apart from dCE itself.

ITQ 11 (*Objective 6*) 1 The lack of interaction between *D* and *E* suggests that the loci *D* and *E* do not usually work as a unit.

2 The existence of *D*– – is most reasonably explained as a deletion. It is easier to accept a single deletion as the cause of *D*–– than that two deletions are necessary to explain a gene order –*D*–.

3 The existence of *D*– – shows that whatever determines the rhesus antigens is divisible into separate functional units.

ITQ 12 (*Objective 7*) Among the mutants you may have suggested would be:

1 Mutants that lack or have defective flagella and therefore cannot propel themselves.

2 Mutants that lack glucoreceptors.

3 Mutants that are unable to metabolize glucose.

In fact mutants of type 3 will still move towards high concentrations of glucose as a result of the responses of their glucoreceptors and flagella even if they cannot utilize the sugar once they arrive at it. Mutants of classes 1 and 2 would generally be classed as behavioural mutants.

ITQ 13 (*Objective 7*) Superficially interpretation (i) might seem reasonable, and the ancillary evidence might also have led you to think that (ii) was possible and (iii) probable. If you chose those interpretations, you would be in company with a number of researchers who have in the past (and to their cost!) done the same. But the evidence given is not enough to justify this conclusion: all we can say is that strain A learns this *particular* maze faster than strain B. There may be many reasons for this. For instance, if strain A had a lower threshold to pain or sensitivity to footpad-shock than strain B, then the incentive for it to learn the escape route in the maze would be higher than for strain B; but this would not tell us anything about its 'general learning capacity'. So far as conclusion (ii) is concerned, the high level of acetylcholinesterase might, for instance, follow from a higher rate of discharge of nerve fibres from the footpads rather than being a 'cause' of learning. At best, one could talk of a correlation. To go further, it would be necessary to show that the levels of no other enzymes were changed, and that the level of acetylcholinesterase changed only when learning occurred. Conclusion (iii) cannot be justified by a single cross of the type described: you should now be aware that F_2 and backcross progeny would need to be analysed before we can make sensible statements about the heritability of character. In fact, interpretation (iv) would seem the only reasonable one to make at this stage.

ITQ 14 (*Objective 7*) By setting the animals from the two strains other learning tasks not involving shock or mazes (for instance, learning to press a lever for a food reward or to recognize light cues) and seeing how they perform on these new tasks. This type of study might indicate whether one was dealing with a very general genetic effect on learning behaviour, or one that was specific to the task involved.

ITQ 15 (*Objective 2*) This is due to the ascertainment problem (Section 14.1.1). In fact, PKU is transmitted as a simple Mendelian recessive character.

ITQ 16 (*Objective 2*) It conforms to the pattern expected for Mendelian inheritance of a dominant allele.

ITQ 17 (*Objective 8*) You should have thought of at least two. Working-class people might be diagnosed more frequently as schizophrenics than middle-class people because middle-class people go to the doctor before the full development of symptoms, because medical consultation regarding 'mental' as opposed to 'physical' symptoms is not equally distributed among social classes, or middle-class people may display their 'eccentricity' in more acceptable ways. Alternatively, people with a tendency to become schizophrenic or with a history of schizophrenic illness may not be able to hold down 'middle-class' jobs and hence will tend to 'drift' downwards socially and live in city slums. This is yet another example of the hazards of 'causal' interpretation in the area of behaviour; but this time we have an example that affects 'environmental' causations more than 'genetic' ones.

ITQ 18 (*Objective 1*) A simple method would be to measure differences in the electrophoretic migration rate of known proteins. Changes in the net charge of the protein can be caused by mutations that alter particular amino acids in the protein. A change in net charge will affect the rate at which the protein moves during electrophoresis. This method was described in Units 9 and 10, Section 10.6.4.

ITQ 19 (*Objective 9*) (a) (i) and (iv)
(b) (ii), (iii), (v), (vi) and (vii)
(c) none

None of the events in (b) will result in changes in allele frequency unless it can be shown that some genotypes are selectively added to or subtracted from the population. Assortative mating (v), recombination (vi) and inbreeding (vii) may lead to changes if these events form homozygotes, which are selected against (i); otherwise they have no effect.

ITQ 20 (*Objective 10*) Although for the MN blood-group system the two populations have very similar allele frequencies, the frequencies of the A, B and O forms of the other blood group show that a distinct difference exists between the populations. In this case, the difference is as striking as the complete absence of the B form in the Navajo sample. So, we conclude that the two populations are not very closely related.

Had we looked only at the MN blood-group system we should have drawn one conclusion, and had we looked only at the ABO system we should have come to another.

ITQ 21 (*Objectives 10 and 11*) (iii) is by far the most reasonable explanation. It seems probable that the common ancestor brought to the area both culture (language, form and customs) and genes (the composition of the gene pool).

Conclusion (i) neglects the arguments that were made earlier in these Units about confounding a common environment (taught language) with common genes.

Conclusion (ii) implies that genetic differences affecting language skill had significant selective value and thus affected their relative frequency—an unlikely but nevertheless possible explanation!

ITQ 22 (*Objective 11*) Explanation (ii) is more likely.

It would be very unreasonable to suggest that the environment was so different in the new continent that generally deleterious phenotypes were favoured only in this unusual community and not elsewhere in the USA. On the other hand, sampling variation could have a strong impact on the allele frequency in a small, initially isolated, community, so one accepts explanation (ii), of genetic drift or sampling variation.

ITQ 23 (*Objective 11*) The factors include those that were mentioned in ITQ 22, namely selection, mutation and genetic drift. Genetic drift is rather less likely if the populations are large, but may be significant in this example if the original West African Negro slave population was a small sample.

ITQ 24 (*Objectives 10 and 11*) Only conclusion (i) can be eliminated entirely. On our definition of a race as a group sharing a common and distinct gene pool, each separate Jewish group would be considered to be a different race.

Conclusion (ii) cannot be refuted or accepted on the basis of the data presented here and would require the demonstration that selection had indeed been operating on this particular allele.

Conclusion (iii) is in agreement with the data and in a sense is confirmed by an analysis of the allele frequency in the Roman Jewish ghetto, which really was very definitely isolated from the rest of Italy from 1554 to 1870. In this instance there *are* differences in allele frequencies between Roman Jew and the non-Jewish Italians.

ITQ 25 (*Objective 12*) Such a conclusion is no more valid than that on the basis of the Stanford–Binet results that middle-class children are 'really' more intelligent than working-class children. All that is really 'proved' is that tests in such studies test what they are constructed to test.

ITQ 26 (*Objectives 12, 15 and 16*) We would say that only a rather simple interpretation of 'the environment' was answered by such a control. The Black population in the USA is part of a minority group in a majority culture that enslaved its ancestors and now discriminates against it on the grounds of skin colour. This represents a cultural difference that cannot be equilibrated by simple manipulation of 'socioeconomic status'. American Blacks living in Northern US cities do tend to score higher than Southern Blacks. On the other hand, some other minority groups, such as American Indians tend to score higher than Blacks on the so-called 'culture-fair' tests. Whether a test is seen as 'culture-fair' or not seems to depend to some considerable extent on the social and cultural interpretation of the test-designer; that is, on what he or she *regards* as cultural universals.

ITQ 27 (*Objectives 15 and 16*) Even if more blood group A individuals than O or B individuals died of stomach cancer, this would only affect the frequencies of the alleles in the next generation if death occurred before reproduction, that is, before these people had completed their family.

ITQ 28 (*Objectives 15 and 16*) The relative fitness of the dwarfs is given by the ratio of their fertility to that of their normal sibs, namely 0.25/1.25 or 0.20.

(a) The selective coefficient, (S) was therefore $1.00 - 0.20 = 0.80$, indicating that there was strong selection at the locus for achondroplasia.

(b) Because selection is operating strongly against a dominant allele one might expect that the frequency of the allele would drop.

(c) If the dominant allele for dwarfism were being removed from the population as a result of the clearly reduced fertility of dwarfs, then the frequency could only remain constant if new alleles were being introduced into the population by mutation.

ITQ 29 (*Objectives 15 and 16*) The mutation rate is 43 per 10^6 gametes. The 94 075 babies had two copies of this autosomal gene each as they are, of course, diploid. The total number of gametes produced in the formation of these zygotes is therefore $2 \times 94\,075 = 188\,150$. As the mutant allele is dominant, each of the 'dwarf' babies need have only one copy of it, but two of these ten mutant alleles are likely to have come from the achondroplasic parent, so only eight are likely to have arisen by mutation in the parents' gametes. Thus the frequency of mutation is 8 per 188 150 gametes or 43 per 10^6 gametes.

ITQ 30 (*Objectives 10, 15 and 16*) The allele frequency would increase. Since the equilibrium allele frequency q_c is given by the following relationship between the mutation rate, μ, and the coefficient of selection, S

$$q_c = \sqrt{\left(\frac{\mu}{S}\right)}$$

an increase in the mutation rate will result in an increase in the value of q_c.

ITQ 31 (*Objective 16*) Inbreeding is mating between related individuals, and leads to the appearance of homozygotes. If the degree of inbreeding decreases, the proportion of homozygotes formed will also decrease.

ITQ 32 (*Objective 16*) The variance for height will increase. Genotypic assortative mating will generate homozygotes that determine the extreme phenotypes. These will make large contributions to the variance, as they will be furthest from the mean: they give rise to the tallest and the shortest phenotypes. Heterozygotes will generate more heterozygotes.

ITQ 33 (*Objective 16*) The heritability of the character would have to be known. If no genetic variation were manifest as phenotypic differences affecting the character, then no amount of phenotypic assortative mating would produce any change, as the differences would be environmental and not genetic.

ITQ 34 (*Objectives 16 and 17*) (a) 0.25

(b) nearly zero

(c) 0.50

(a) A phenylketonuric is a homozygous recessive genotype. Both parents must, therefore, have contributed a mutant allele (unless the child itself was a mutant) but as neither showed the effect, they themselves must be heterozygotes. Thus the probability of re-occurrence is the product of the probabilities of each parent producing a gamete carrying the recessive allele, which is $0.5 \times 0.5 = 0.25$.

(b) As dwarfism appeared in neither parent even though it is due to a dominant mutation of good penetrance (that is, it always manifests itself phenotypically when present in the genotype) then we must presume that the dominant allele arose as a mutation in the gonads of one or other parent. In this case the probability of recurrence of the same mutation is infinitely small.

(c) The brachydactylous child was born to a parent also carrying that mutation. If we assume that the affected parent is a heterozygote (more likely than a homozygote, even if it were a viable condition) then the probability of transmission is 0.5.

Answers to SAQs

SAQ 1 (*Objective 5*) (a) Heterokaryons are fused cells that have cytoplasm in common but distinct nuclei. In hybrids, nuclear fusion has also occurred.

(b) Heterokaryons—gene–cytoplasm interactions, complementation.
Hybrids—mapping, complementation, linkage.

SAQ 2 (*Objective 9*) 1 A, 2 A, 3 B, 4 A, 5 B, 6 C, 7 C

SAQ 3 (*Objective 12*) Originally it was the ratio of a child's 'mental age' as measured on a standard test, to his or her chronological age; today, it is based on the ratio of an individual's score on a standard test to the population mean for that test, which is adjusted to 100 (Section 15.3.2).

SAQ 4 (*Objective 12*) The principal reasons given in these Units (Section 15.3.2) are that:

1 Intelligence is not a global character but a compressed way of referring to a whole range of skills and abilities.

2 Intelligence is not a fixed and stable attribute of an individual but varies during development and later life.

3 The test measures that are used select out certain socially approved criteria for scoring, and are inevitably biased towards a majority population.

4 An individual's test score varies according to the nature of the test and how it is administered. It is thus the product of a relationship rather than an individual character.

SAQ 5 (*Objectives 5, 8 and 13*) (a) *Schizophrenia* It is difficult to see how 1 or 5 can be used or what could be achieved from attempting their use. 3 might potentially reveal biochemical differences between schizophrenics and non-schizophrenics, though it is not certain what light this might cast on the genetic mode of transmission. 2 and 7 may be and have been used, but they do present problems of the confounding of genetic and environmental effects produced by a common rearing pattern in the same family. 4 and 6 are the most promising approaches, but you will be aware from the text how difficult it is to find enough well-controlled MZ twins reared apart to constitute a sample for study.

(b) *Intelligence* The same arguments apply as in (a), only the problems with approach 3 become even more acute when one is dealing with what is clearly a continuously varying character such as intelligence rather than a character such as schizophrenia where some at any rate would maintain that there is a comparatively clear distinction to be drawn between sufferers and non-sufferers.

SAQ 6 (*Objective 14*) You should have recognized that because heritability is a within-population statistic no experiment could measure the heritability of *between-population* differences. On the other hand, if intermarriage were to result in the merging of Whites and Blacks into a single interbreeding group (converting them to one race, in the genetic sense) then the heritability of differences in height could be measured.

SAQ 7 (*Objective 18*) See Section 15.4.5. The arguments against are:

1 It is unlikely that all the 'desirable' characteristics of the individual are genotypically determined.

2 It is unlikely that they are exclusively carried in the male genotype, of which the female is a passive carrier.

3 That the characteristics that make an individual 'great' in one context will be similarly perceived in another context is improbable.

There are no arguments in favour that we can think of.

SAQ 8 (*Objectives 16, 17 and 18*) In our view the following are the most likely answers to this question:
1 D, 2 B, 3 A, 4 C, 5 D

Bibliography and references

1 General reading

Cavalli-Sforza, L., and Bodmer, W. F. (1971) *The Genetics of Human Populations*, Freeman. (A complete and definitive advanced text.)

Ebling, F. (ed.) (1975) *Racial Variation in Man*, Blackwell.

Jencks, C. (1973) *Inequality* (Boards) Allen Lane; (Paperback) Harper & Row.

Jones, A., and Bodmer, W. F. (1974) *Our Future Inheritance: Choice or Chance?*, Oxford University Press.

Kamin, L. J. (1975) *The Science and Politics of IQ*, Erlbaum.

Lerner, I. M. (1968) *Heredity, Evolution and Society*, Freeman.

McClearn, G. E., and De Fries, J. C. (1974) *Introduction to Behavioural Genetics*, Freeman.

McKusick, V. A. (1969) *Human Genetics*, 2nd edn, Prentice-Hall. (An excellent and interesting introduction.)

Richardson, K., and Spears, D. (eds) (1972) *Race, Culture and Intelligence*, Penguin.

2 Other publications cited in the text

Clarke, C. A. (1968) The prevention of 'Rhesus' babies. *Scient. Am.*, November, 46–52.

Edwards, A. W. F., and Cavalli-Sforza, L. L. (1964) 'Reconstruction of evolutionary trees' in *Phenetic and Phylogenetic Classification* by V. E. Heywood and J. McNeill (eds) Systematics Association Publication no. 6, The Systematics Association.

Erlenmeyer-Kimling, L., and Jarvik, L. F. (1963) Genetics and intelligence: a review. *Science, N.Y.*, **142**, 1477–9.

Etzioni, A. (1975) *Genetic Fix: The Next Technological Revolution*, Harper & Row.

Eysenck, H. J. (1971) *Race, Intelligence and Education*, Temple Smith.

Feldman, M. W., and Lewontin, R. C. (1975) The heritability hang-up. *Science, N.Y.*, **190**, 1163–8.

Galton, F. (1869) *Hereditary Genius*, Macmillan.

Harrington, G. M. (1975) Intelligence tests may favour majority groups in a population, *Nature*, **258**, 708–9.

Holland, H. W. (1883) *Atlantic Monthly*, **52**, 477, quoted in *Eugenics and the Progressives* by D. K. Pickens, Vanderbilt Press (1968).

Jensen, A. R. (1969) How much can we boost IQ and scholastic achievement? *Harv. educ. Rev.*, **39**, 1–123.

Landsteiner, K. and Wiener, A. S. (1940) An agglutinable factor in human blood recognised by immune serums for Rhesus blood, *Proc. Soc. Exptl. Biol. Med.*, **43**, 223.

Levine, P. and Stetson, R. E. (1939) An unusual case of intra-group agglutination, *J. Amer. Med. Ass.*, **113**, 126.

Montagu, A. (1972) *Statement on Race*, Oxford University Press.

Mourant, A. E., Kopec, A. G., and Domaniewska-Sobczak, K. (1976) *The Distribution of the Human Blood Groups and Other Polymerisms*, 2nd edn, Oxford University Press.

Muller, H. J. (1936a) Unpublished autobiography.

Muller, H. J. (1936b) *Out of the Night*, Vanguard Press.

Osborne, R. H., and de George, F. V. (1959) *Genetic Basis of Morphological Variation*, Harvard University Press.

Pickens, D. K. (1968) *Eugenics and the Progressives*, Vanderbilt Press.

Pontecorvo, G. (1961) Genetic analysis via somatic cells. In 'The Scientific Basis of Medicine', *Annual Reviews, British Postgraduate Medical Federation*, Athlone Press, 1961.

Race, R. R., Mourant, A. E., Lawler, S. D., and Sanger, R. (1948) The Rh chromosome frequencies in England. *Blood*, **3**, 689–95.

Race, R. R., and Sanger, R. (1975) *Blood Groups in Man*, 6th edn, Blackwell.

Roberts, J. A. F. (1942) Blood-group frequencies in North Wales, *Ann. Eugen*, **11**, 297.

Rose, S. P. R. (1975) 'Scientific racism and ideology' in *Racial Variation in Man* (ed. F. J. Ebling), instutute of Biology Symposium, No. 22, Blackwell.

Roslansky, J. D. (ed.) (1966) *Genetics and the Future of Man*, North Holland.

Shields, J. (1962) *Monozygolic Twins Brought Up Apart and Brought Up Together*, Oxford University Press.

Schwartz, M., and Schwartz, J. (1974) Evidence against a genetical component to performance on I.Q. tests, *Nature*, **248**, 84–5.

Shockley, W. (1972) Dysgenics, geneticity and raceology, *Phi Delta Kappan*, **53**, 297.

Stern, C. (1973) *Principles of Human Genetics*, 3rd edn, Freeman.

Terman, L. (1924) The conservation of talent. *School and Society*, March, 363; quoted in K. Ludmerer *Genetics and American Society*. Johns Hopkins Press (1972).

Thorndike, E. L. (1927) *Educational Psychology*, quoted in Pickens, *op. cit.*

Acknowledgements

Grateful acknowledgement is made to the following for material used in these Units:

Appendix

M. W. Feldman and R. C. Lewontin, 'The heritability hang-up' in *Science*, **190**, 19 Dec., 1975. Copyright © 1975 by the American Association for the Advancement of Science.

Tables

Tables 1 and 9 from C. Stern, *Principles of Human Genetics*, 3rd edn, W. H. Freeman and Co. Copyright © 1973; *Table 3* based on data from R. H. Osborne and F. V. de George, *Genetic Basis of Morphological Variation*, 1959, Harvard University Press; *Table 10* from R. R. Race *et al.*, 'The Rh chromosome frequencies in England' in *Blood*, **3**, 1948, Grune and Stratton Inc. by permission; *Table 15* from A. E. Mourant *et al.*, *The Distribution of Human Blood Groups and other Polymorphisms*, 2nd edn, Oxford University Press 1976.

Figures

Figures 6, 7 and 17 from C. Stern, *Principles of Human Genetics*, 3rd edn, W. H. Freeman and Co. Copyright © 1973; *Figure 10* from B. Ephrussi and M. C. Weiss, 'Hybrid somatic cells' in *Scientific American*, **220**, No. 4, April, 1969; *Figure 14* from A. W. F. Edwards and L. L. Cavalli-Sforza, 'Reconstruction of evolutionary trees' in *Phenetic and Phylogenetic Classification*, Heywood and McNeil (eds), 1964 Systematics Association; *Figure 15* from A. E. Mourant *et al.*, *The Distribution of Human Blood Groups and other Polymorphisms*, 2nd edn, Oxford University Press 1976; *Figure 16* from L. Erlenmeyer-Kimling *et al.*, 'Genetics and intelligence: a review', in *Science*, **142**, 13 Dec., 1963. Copyright © 1963 by the American Association for the Advancement of Science.

Swansea College of Further Education

1   What is Genetics?

2   Chromosomes and Genes

3
     Recombination, Linkage and Maps
4

5   Chromosomes: Organization and Changes

6   Molecular Genetics

7   Cytoplasmic Inheritance

8   Developmental Genetics

9
     Analysis of Populations
10

11   Biometrical Genetics

12   Theories of Plant and Animal Breeding

13   Ecological and Evolutionary Genetics

14
     Human Genetics
15

HIST   The History and Social Relations of Genetics

STATS   Statistics for Genetics

LC   Life Cycles